Rescues and Disasters in Western Canada

CANADIAN ART GALLER
LESALE

HELP!

by Faye Reineberg Holt

Rescues and Disasters in Western Canada

Altitude Publishing
Canadian Rockies/Vancouver

Cover: *Mountain Rescue Training School has been offered in the Rocky Mountains since 1954. Originally, the school had two general objectives. Courses taught park wardens and RCMP how to rescue victims after mountain and avalanche accidents. They also taught trainees how to reach and fight mountain fires. This student has volunteered to be a victim while the other students bring him safely down the mountain.*

Frontispiece: *A September 1, 1951 fire at the Cummings Block in Calgary trapped Mrs. Clara Litovchenko and four other people who were rescued by firemen.*

Back Cover: *When a railway sleeping car derailed and fell to the ice of the North Saskatchewan River in Saskatoon, RCMP, firemen and rescuers faced the challenge as a team. See March 6, 1912 for story.*

Publication Information

Canadian Cataloguing in Publication Data
Reineberg Holt, Faye.
Help!
ISBN 1-55153-909-8
1. Rescue work–Canada. 2. Natural disasters–Canada. I. Title.
GB5011.15.R44 1997 363.3'48'0971 C96-910868-0

Art Direction: Stephen Hutchings
Design: Lisa Belter
Electronic page layout: Lisa Belter
Editor: Sabrina Grobler
Project management: Sharon Komori
Sales management: Scott Davidson
Financial management: Laurie Smith

Made in Western Canada
Printed and bound in Canada
by Friesen Printers, Altona, Manitoba.

Altitude GreenTree Program
Altitude Publishing will plant in Canada twice as many trees as were used in the manufacturing of this book.

9 8 7 6 5 4 3 2

Altitude Publishing Canada Ltd.
1500 Railway Avenue, Canmore, Alberta
Canada T1W 1P6

Contents

Gold Rush Town in Ashes

Townspeople Risk their Own Lives

SEPTEMBER 17, 1868

Barkerville, BC — People were on their own, fighting the devastating blaze of September 16 in this isolated northern community. Despite living in conditions under which fire poses a very real threat, this gold mining community has no fire brigade. Instead, it was ordinary townspeople who risked their own safety to prevent loss of life. Those individual acts of courage are credited with saving everyone from being blown sky high.

One store had 50 kegs of blasting powder for sale. With the building in the path of the fire, quick-thinking individuals carried the explosive to a nearby dry shaft. They stored the powder at the bottom of the shaft and prevented an all-out holocaust. Had the explosives ignited, it is unlikely that there would have been any survivors. Even the small coal-oil tins that exploded were capable of blasting blankets and bedding 200 feet [61 m] into the air. After the fire, the top of an exploded tin was discovered five miles away.

The firestorm, which raged through the townsite and up to the mountain ridge behind it, left thousands homeless. Unfortunately, there are no nearby neighbors to take victims in from the cold. They must fend for themselves, and last night, most slept under the stars.

Driven by high winds, the fire spread through town at a frightening speed. The town's wood-frame buildings and its narrow main street meant that there was no stopping the sheets of flame. For firefighting, the only nearby water source is the Barker Flume, 50 feet [15 m] overhead. Standing on the flume and using its water, some men were able to delay the spread of the fire to the upper end of town. Unfortunately, the intense heat made it impossible to continue. Although the fire lasted only two hours, it forced people to abandon their efforts to save property. The only refuge was the middle of Williams Creek.

The fire destroyed 116 homes and many other buildings. Only warehouses at the far end of town, shacks in Chinatown and Scott's Saloon near the flume are left standing. Damage is expected to be in excess of $675,000.

The frosty winds of winter are funneling down the canyon, and shelter is of serious concern. The nearby sawmill was not damaged, and today, people are beginning to rebuild in spite of the fact that the price of lumber

has roared from $80 to $125 per thousand board feet.

The fire occurred at the end of a summer of drought. It started at 2:45 p.m. in Barry and Adlers' saloon. One of the girls was ironing in the kitchen. When a miner tried to kiss her, their struggle dislodged a stovepipe and the canvas ceiling caught fire. Flames immediately engulfed the roof.

The town was established in 1862 after English seaman Billy Barker rediscovered gold on Williams Creek. By that time, the Caribou gold fields had experienced a boom and bust cycle. At one time, prospectors had successfully panned on Williams Creek. When easy riches disappeared, miners began to dig and drive shafts in other areas. They assumed that Richfield Canyon was the natural sluice and that there would be no gold beyond it. Barker proved them wrong. He drove a shaft on his 600-foot [183 m]-long claim, and at 52 feet [16 m], he discovered gold. Intent on making their fortunes, thousands rushed back to the area. ■

Photo

The Barkerville fire has created a crisis of homelessness. Because of the gold rush, it was once the largest town north of San Francisco and west of Chicago. Now, only one house remains standing in the residential and business sections.

Captain Ignores Screams for Help

Frigid Water Claims Hundreds

NOVEMBER 10, 1875

Victoria, BC — Almost a week after the November 4 disaster, when the S.S. *Pacific* sank, it seems certain that the only survivors are one passenger and one crew member. As many as 300 have lost their lives. The majority of the victims on the giant American ship were residents of Victoria. At the city's docks, families still gather in the hope that the next ship will have their loved ones aboard. Unfortunately, the frigid waters of the Pacific Ocean make it virtually impossible for passengers to have survived.

After sailing through the Strait of Juan de Fuca, the S.S. *Pacific* was struck by the 1,100-ton [1,117,600 kg] *Orpheus*. *Orpheus* sailors have charged their captain with ignoring screams for help from people on the passenger ship. Some claim that Captain Sawyer dismissed their pleas in order to lower a boat to check on the safety of those aboard the *Pacific*. They also say that he often drank spirits. The captain has denied that he heard any screams. However, he has been arrested, and an American inquiry is being launched. Sawyer also faces charges of purposefully beaching the *Orpheus* after the disaster.

The two survivors of the disaster were rescued by American ships. Henry Jelly was found at 10:00 a.m. Saturday morning. The Ontario-born passenger had tied himself to the wheelhouse. He floated for two days until he was discovered within three miles of Vancouver Island. Crewman Neil Henley floated on the debris from Thursday until his rescue on Monday morning.

The doomed S.S. *Pacific* had been traveling from Victoria to San Francisco. Most Victoria families had at least one relative on board. However, the exact number of fatalities may never be known. During the price wars, some passengers were allowed to travel for free, so their presence goes unrecorded. As well, children are not ticketed; only when a family reports them missing are they considered to be among those lost.

As the two men who were rescued begin to tell their stories, Victoria families are becoming even more distressed. Apparently, the *Pacific*'s five lifeboats had a maximum capacity of only 160 people. To make matters worse, there had been no lifeboat drill. In fact, after leaving port, some boats were filled with water to control the ship's listing, probably from improperly

stored cargo.

After the collision, confusion reigned. Lifeboats were drained, but some plugs were not replaced. The mechanism for lowering the boats jammed. The crew used axes to lower some, while others remained raised, as if the crew presumed that they would float once the ship sank. Two boats had no oars. Survivors suggest that more time was lost when disagreements flared over who would man the boats carrying women and children.

After the ship broke in two, panicking victims swamped the lifeboats. In their voluminous dresses, women were doomed. Twenty people managed to cling to wreckage, but eventually, the freezing water took their lives.

Bad weather and rough seas are not considered to have been factors in the accident. Three men were assigned to the watch on the *Pacific*, but they must not have seen the *Orpheus*. After the collision, lights went out. Incoming water extinguished the engine boilers and ruptured the *Pacific*.

Debris reveals that the ship was plagued with dry rot and rust. The American vessel was not subject to Canadian inspections. While it was to dock in Canada and was carrying Canadians, it was only subject to inspection if it carried mail. Throughout North America, inspections are rare and generally superficial. If a ship is scheduled to be inspected, crew often borrow equipment from another ship until the inspector leaves. Other factors suspected of having contributed to the disaster include the number, training, discipline and workload of the crew and officers. ▪

Photo

The American side-wheeler, the S.S. Pacific, *was 24 years old when it sank. During its career, it plied routes along the Pacific northwest coast from San Francisco to Alaska.*

Fear Sweeps the West

NWMP, Army, Volunteers Remain on Alert

AUGUST 1, 1885

Regina, SK — After months of fear on the part of settlers in Western Canada, the Native and Metis rebellion has ended. Today, leader Louis Riel was sentenced to hang for treason. Other rebel leaders have also been imprisoned. No further violence is expected, but some settlers are still anxious. Therefore, the North West Mounted Police, General F. Middleton's troops and the volunteer militia remain vigilant.

A general fear swept the West when Metis rebels established a provisional government at Batoche on March 19, 1885. When the Canadian government realized that the Metis declaration jeopardized the unity of the nation, it declared the action treasonous. By late March, when settlers felt threatened and at risk, the government mobilized the army. Militia units from Winnipeg and Eastern Canada traveled by train to the larger communities in the far west. Their duty was to assist the North West Mounted Police and local volunteers in quashing the rebellion.

After the Frog Lake Massacre, when Metis and Natives took two women captive and nine men were killed, panic escalated. In other areas, property was looted and ransacked, but only a few other civilians died or were physically hurt. Still, many frightened settlers moved to the NWMP forts for protection.

The presence of the military helped calm fears. Under the command of General F. Middleton, almost 8,000 men prepared for battle. In this first major military action in Western Canada, soldiers faced fearless and well-organized Natives and Metis. Lives were lost on both sides. At first, it was the rebels who claimed significant victories. On May 12, the tide turned.

After three days of fighting, Batoche fell to Middleton's forces and weaponry, which included the Gatling gun. Riel and members of his rebel council were taken to Regina and jailed. Gabriel Dumont, military commander of the Metis forces, has escaped. He is likely in the USA. Other Natives and Metis have fled north.

On May 26 at Battleford, Chief Poundmaker and his followers surrendered to Middleton. To the north, hundreds searched for Chief Big Bear. On July 2, rather than escape to the USA, Big Bear surrendered at Fort Carlton. Compounded with the conviction of Riel, the

Government of Canada has again established itself as the valiant representative of law and order in Western Canada. The job of enforcing those laws is once again in the hands of the NWMP.

The Northwest Rebellion appears to have resulted from a lengthy history of land-claim disputes. The aboriginal and Metis people are at odds with settlers and the Canadian government over land ownership and the right of Natives and Metis to govern themselves. Those problems were evident in 1869-70 when dominion surveys were interrupted by Metis protesters. Concerned about control of Metis lands, leaders Riel and Dumont declared independence from Canada. When this first rebellion was quashed, both leaders fled to the USA.

After years in exile, Riel and Dumont returned to the Canadian frontier. By the spring of 1885, animosities had escalated. The first rebel action was to seize stores, cut telegraph lines and take prisoners at Batoche. Claiming that conditions on reserves were deplorable and starvation was epidemic, discontented Natives supported the Metis.

The rebellion made panic-stricken western Canadian settlers welcome the army's intervention. However, most do not condone the confrontational approach that became commonplace between settlers and aboriginal people in the settlement of the American West. ■

Photo

These artillery soldiers from the Montreal Garrison form a diamond on the wide open prairies. On April 20, 1885, this battalion from Winnipeg and two militia groups left Calgary for Edmonton as part of General T. B. Strange's contingent. The troops proceeded down the North Saskatchewan River to Fort Pitt. There and at Battleford, fearful settlers had moved to the NWMP forts.

Firemen Blaze New Trails

Police and Fire Services on the Leading Edge

JANUARY 1, 1886

Victoria, BC — The capital of British Columbia has scored another first in providing police and fire services. With the new year, Victoria's volunteer fire brigades are being reorganized into a new style of fire department comprised of 26 full-time firemen. The full-time appointments recognize the essential nature of the service offered during fires. In 1882, the city appointed C.J. Phillips as the fire department's first paid chief. Now, after two decades of service from volunteer firemen, the city is assuming full responsibility for the department. The move makes firefighters in Victoria the first in Western Canada to earn a full pay cheque. The city was also the first in the West to use a steam engine in fire fighting. The Hudson Bay Company, the government and private individuals contributed money to buy the engine, which arrived from New York via San Francisco in 1868.

Police service has been an equally long-standing priority. Before 1858, in a tiny community of about 400 people at Fort Victoria, a fire alarm brought out whoever might be affected or interested. Justice was in the hands of the Hudson Bay Company Chief Factor or officials. By 1852, a jury had assembled in the first courtroom west of Manitoba. The room was on a ship called *Beaver*. Two natives were sentenced to be hanged for the murder of a shepherd, and with that, settlement law was clearly taking shape.

By the end of 1858, when the first shipload of 450 gold rush miners from San Francisco arrived in Fort Victoria, new approaches were needed. The sudden influx of 20,000-30,000 people meant good times, but also problems and change. Wooden shacks were being hastily built and created a fire hazard. Volunteer fire brigades became essential. As well, Chief Factor Douglas wanted the newcomers to abide by British law.

The Americans had brought guns and a rowdy lawlessness. Although he was without authority once the miners reached the mainland, Douglas set up the first official police force to patrol the colony of Vancouver Island. He appointed the first Commissioner of Police and a Police

Photo

In 1882, C.J. Phillips became Victoria's first paid fire chief. He worked part-time and earned $700 a year. The assistant chief was paid $300. At the time, the only full-time worker was the steward, who was paid $900 to look after the fire hall, equipment and horses.

Superintendent. He also appointed a dozen constables, but they were not to bear arms. One month later, he banned the belt gun.

At Fort Langley on the mainland colony of British Columbia, the same problems arose. Therefore, in the same year, the British government appointed a Commissioner of Police, justices of the peace and B.C.'s first judge. Judge Matthew Begbie roamed the interior, acting as judge, prosecution and defense. He was soon known as Canada's hanging judge. The strong arm of the law was alive and well in the far west; never has an area of North America been settled with so little crime, despite having only one policeman to every 4,000 square miles [10,360 km^2].

In the end, many of the American gold rushers who had made the colonies face issues of law, order and fire safety stayed in Victoria and on the lower mainland. They strongly influenced how volunteer fire brigades evolved. Now, with full-time paid firefighters, British Columbia enters a new era in providing essential services. And Victoria is its leader. ▪

Inferno Destroys Vancouver

Firemen Helpless

JUNE 14, 1886

Vancouver, BC — Yesterday, individuals in this coastal community of 3,000 fought their way into flaming buildings to recover belongings and save others. As the inferno raged through the streets, burning timbers fell, high winds created walls of flame and clothing caught fire. Firemen exerted a valiant effort but could do nothing to stop the fire. Only the heroism of others saved many of the townspeople.

A bartender pulled a woman and her child from a fiery building. Local fishermen became rescuers, taking victims from the inferno to ships in the harbour, where they could find refuge. Once the billowing smoke—which was visible for 50 miles [80 km]—signaled the catastrophe to New Westminster and Moodieville, more help arrived. At Moodieville, the Masonic Hall was designated a temporary shelter.

Twenty people have been confirmed dead. Fatality counts will likely rise once they include estimates of newcomers and those individuals without relatives to report them missing.

The bells of St. James Church raised the alarm, but the new volunteer fire department faced the firestorm with only a man-drawn hose reel as equipment. For twenty minutes, buildings appeared to melt. Then winds began to die and the worst of the fire subsided.

During that time, simply surviving required acts of courage. Desperately trying to save themselves, some dropped to the bottom of their water wells. When the fire's intensity sucked out the air, they suffocated. Cut off by the flames, others crawled into tree holes and covered themselves with dirt. Those who found enough water in ditches covered themselves with wet blankets.

False Creek, the CPR embankment and the Westminster Road became refuges for some, but the ocean was the only avenue of certain escape. Many who did not have access to their own boats tied logs together as makeshift rafts.

Unfortunately, as yesterday was a Sunday, many families were separated because of holiday recreations. As a result, many of the survivors are still in agony as they wait for news of loved ones.

Before the fire, Vancouver boasted roughly 800 buildings. Now, only four houses near the False Creek Bridge remain standing. The Regina

Hotel was also saved by men who fought flames with wet blankets and towels.

The blaze started around 2:00 p.m. High winds carried the forest-clearing fire on a nearby hillside to a shed. Huge slash piles of brush, created during the logging and clearing of the area, were tinder dry after the long, hot summer. With ready fuel and winds fanning the flames, the fire quickly raged beyond control.

Originally called Granville and unofficially known as Gastown, Vancouver was incorporated as a city in January of this year, only six months prior to the fire. Already, Mayor Maclean is planning to solicit at least $5,000 from Canada's prime minister. He also hopes to garner a loan of about $10,000 from the Bank of British Columbia for rebuilding. ■

Photo

The morning after the Vancouver fire, there are a few signs of hope. Tents have been erected for the homeless. To deal with the aftermath, City Hall has been relocated to a tent. There, victims have access to police, fire service and city administration.

Sick Mounties Suffer

Hospitals Mean Hope

DECEMBER 26, 1888

Regina, SK — The annual report submitted to the North West Mounted Police Commissioner tells of Mounties who travel for days across the prairies to reach a hospital. These dedicated men are responsible for law and order in the West. They endanger themselves confronting criminals and drunks. Mounties sit "death watch" with prisoners who are to be hanged. They watch for fires and fight them when necessary. They ride wherever they are needed to check on isolated settlers and prospectors. They pitch in to help the sick when they arrive at homesteads where families are ill. And like everyone else, sometimes they succumb to sickness.

Despite having been in the far west for almost 15 years, seriously ill Mounties sometimes travel from as far as the mountains of the Alberta territory all the way to Regina for adequate hospital care. Hypo-malaria fever is one of the worst threats to their health. This past year, three such patients traveled to Regina from the Wood Mountain Post near the American border in the Saskatchewan territory. There, since the creek that supplies water to the post is used for drainage from some settlers' houses, epidemics are to be expected, but adequate treatment is not nearby.

Elsewhere, problems are similar. The assistant surgeon for Fort Macleod cares for two divisions. Among the men of D Division in the Crowsnest Pass of British Columbia, he treated a typho-malaria epidemic involving a death. The Mounties at Fort Macleod do have a hospital, but others are not so fortunate.

Prince Albert has no hospital. The assistant surgeon at Fort Saskatchewan reports that his hospital was flattened by fire on January 31, 1888. He visits the Edmonton post weekly. However, one gunshot patient could not be moved. The best the NWMP surgeon could do was to send a hospital sergeant to Edmonton to provide care.

Maple Creek has a building used as a hospital. As well as being too small to serve the number of men in the division, it leaks rain and the wind passes through the walls all year round. Lethbridge just built a hospital, but the assistant surgeon there has no surgical instruments or appliances. Calgary reports that the construction of a hospital will probably be completed by spring.

The assistant surgeons

are not ones to complain, and their reports are filled with thanks for each small improvement. The Mounties themselves will certainly survive with few services and little help. On the long march west, they faced locusts, mosquitoes, hail, heat, saline water, muddy water and inadequate food. Often, they cared for their exhausted and starving horses before they looked after their own needs. Ironically, those who were weakest were sent to Edmonton by way of Fort Ellis, Pitt and Carlton. It was supposed to be an easier route than the southerly one. Their journey proved to be more difficult, and they trudged 800 miles [1,287 km] farther than the other divisions. Over 88 arduous days they cleared paths for their wagons and finally reached Fort Saskatchewan, which has no hospital.

True, the most common health problems for the men are coughs, colds, contusions, diarrhea, coryza, flesh wounds, frost bite and rheumatism. However, if they are to police the West, fight fires and help settlers in trouble, they deserve well-built, adequately equipped hospitals. Most of the men are brave and hardy. Since the Riel Rebellion, few have lost their lives. But surviving without services is almost as challenging as the long march itself. When will they get the help they deserve? ▪

Photo

At Regina where the Mounties have their best hospital facilities, Assistant Surgeon Dodd (standing centre) treated many sick men during 1988. He performed surgeries for fractures, fistula, gunshot wounds and frostbite. One constable remained on duty until two days before his death. For some time, he had suffered from a serious abscess that ultimately caused his death. Dodd reports his amazement that the man was able to do his job while enduring the pain.

Homeless Forced to Flee Flood

Livestock Threatened

MAY 5, 1893

Emerson, MB — Throughout the floodplain, the homeless have been forced to find shelter on higher ground with friends and relatives. Residents near Rosenfeld have come to the rescue of new immigrants from Russia. When they arrived expecting to begin building homes and plowing land, many found that their land was flooded. Fortunately, some immigrants had relatives nearby, and many strangers were welcomed into the homes of those not affected by the flood.

Problems obtaining clean drinking water have put people at risk of typhoid. Electrical outages and the inability to maintain heating and cooking fires cause suffering for everyone. This poses special problems for women who are caring for young children. It also means additional risk to the elderly during the cold, wet weather. To rectify water, electrical and heating problems, skilled and unskilled men are devoting long hours and hard work.

To save 150 young pigs, the manager of a farm took his pigs to the upper storey of a very large granary. He will keep them there unless floodwaters threaten to overturn or carry the building away. For farmers, the task of caring for livestock is unbelievably difficult, and losses have been inevitable. Given adequate warning, farmers have attempted to move their stock in rowboats and on rafts when necessary. Lack of feed further complicates matters, and too often, farmers have been unable to save their stock.

Yesterday in Emerson and Morris, floodwaters were still slowly rising. Finally, the flood plaguing the Red and Assiniboine River basins is expected to subside. In Winnipeg, water levels approached the 1882 high-water mark, but now, the worst is over for that city.

At their worst, floodwaters rose 12 inches [30 cm] in the course of 24 hours from midnight on May 1 to midnight on May 2. For more than a week, the prairie has been covered with two inches [5 cm] to 24 inches [61 cm] of water, and the devastation has been alarming.

Floods are reasonably common here. The Red River Valley drainage area extends over 111,000 square miles [287,468 km^2], 50,000 square miles [129,490 km^2] of which are fed directly by the Red River. The Assiniboine, which flows into the Red, has flooded the rest of the area. Because the land is flat and the river channels

carved onto the glacial plain are not deep, the area floods easily.

Historically, floods have occurred after several years of wet summers. The ground becomes water-logged. Then there is a late spring, heavy snow pack, thick ice on the rivers and a late break-up. Combined with heavy, wet snow flurries and a deluge of rain in the spring, flooding becomes a threat.

Once break-up occurs, ice jam can cause more problems. When drainage is blocked and water is held back, levels that are otherwise kept low by rivers suddenly rise. Once the Red River overflows its banks, the water spreads rapidly because the land is flat.

This spring, conditions were right for a flood, and concern built slowly. The first hint of the impending disaster was on April 20. Minnesota and North Dakota were having serious problems. Because of the earlier river break-up and spring melt, the American experience generally foreshadowed a Canadian problem. As most people believed themselves to be safe, disaster planning came too late. Individuals saved what property they could and will have to clean up the damage, but at least no lives were lost. ▪

Photo

Sidewalks and large buildings have been flooded and smaller ones have been carried away at Emerson, 60 miles south of Winnipeg. Canoes and rowboats are now the only means of rescue and transportation. Because of weakened bridges and washouts, scheduled trains have been temporarily canceled between Winnipeg and the community on the American border, where floodwaters call to mind the Great Lakes. Telegraph lines are also down. Emerson has survived a long history of floods. It is 1-1/2 miles [2.4 km] from the old Boundary Commission post of Fort Dufferin, an abandoned HBC fur-trading post.

Streetcar Plummets from Bridge

Society Women, Boaters Come to the Rescue

MAY 26, 1896

Victoria, BC — About 90 streetcar passengers have been rescued after the centre span of the Point Ellice Bridge, which links Victoria and Esquimalt, collapsed. Twenty-seven of those rescued are seriously injured. Another 55 are dead as a result of the worst streetcar-related incident in North America and the worst bridge disaster in Canada.

Residents from the elegant Point Ellice district and others did everything possible to save victims of the holiday weekend tragedy. Once victims were rowed ashore, the home and lawn of shipping mogul Captain Grant became a nursing station. Mrs. Grant, the Tyrwhitt-Drake family and other residents of the prestigious neighborhood tried to resuscitate victims. They provided blankets, dry clothing and what comforts they could to help the survivors.

The unfortunate fatalities, including many children, also began to line the lawn. While they were still in shock, relatives and friends identified loved ones. Curtains from the Grant home were used as shrouds for the dead.

At the time of the collapse, one bicyclist and two horse-drawn carriages were also on the bridge. The cyclist perished. The carriage drivers and their passengers were saved. Caught in their lines, all the horses drowned.

Holiday crowds aggravated the disaster. Streetcars were overloaded with passengers intent on celebrating the Victoria Day weekend. Yesterday, a community picnic, races and a regatta were staged at the Gorge. Before today's disaster, military maneuvers and a mock battle had been planned for Macauley Point.

Although its capacity was 60 people, Car Sixteen was crowded with passengers to watch the maneuvers and the battle. The 21-ton streetcar was far in excess of the 10-ton capacity for the bridge.

Since it was built in 1885, the four-span bridge has been a source of continual concern. At that time, heavy streetcars were not in use. The first person

Rotting timber is considered a possible cause of today's disaster. ■

Top Photo

After the Point Ellice Bridge collapsed, boaters near the bridge began rescuing those who struggled in the water. Some had been standing on the outside platforms of the streetcar. Others inside Car Sixteen escaped through open windows into the water. Most who became submerged inside the car could not be pulled free. Divers rushed to the scene, but time was not on their side and fatalities soared.

Bottom Photo

Mr. Tyrwhitt-Drake, former mayor of Victoria, lived at Ellice Point. In this gathering are youthful members of his family and members of the Drake family. Two of the mayor's daughters pulled victims out of the water and into their boat. Their rescue efforts saved at least seven people from drowning.

to cross Ellice Point Bridge did so in 1890. In 1893, while a streetcar was crossing, the roadbed on the bridge sagged three feet [0.9 m]. The streetcar made it safely to the other side. The bridge was strengthened, but the incident confirmed the growing uncertainty regarding its load-bearing capacity.

Tragedy Befalls Climber

American First Climbing Fatality in National Parks

AUGUST 5, 1896

Banff National Park, AB — Two days ago, a tragic fall took the life of famed mountaineer Phillip S. Abbot. He was climbing with Charles Fay, Charles Thompson and George Little, all from the United States. Today, his fellow climbers received assistance in recovering the body of their friend. Although they are not climbers themselves, NWMP Inspector Harper, Park Warden Tom Wilson and chalet manager Willoughby Astley accompanied them on the difficult mission.

The death of Abbot on Mount Lefroy devastated alpinists around the world. On August 3, 1896, he fell and struck an ice slope 15 feet [4.6 m] below his fellow climbers. Once his fall was broken, Abbot rolled down the steep incline. His companions immediately began their rescue attempt. Without a rope, it took them more than three grueling hours to reach him.

Unconscious but alive, Abbot was carried only a short distance before he passed away. Given the difficult return and their physical exhaustion, the group left his body. By 10:30 that night, the surviving climbers had descended only as far as the Death Trap Col. The col was named by Samuel Allen in 1884 and was the summit of the pass between Mount Victoria and Mount Lefroy. There, the climbers spent the night.

Despite light snow and wind, they had to go still further down the mountain the next morning to retrieve their rucksacks and eat for the first time in 16 hours. In the rain, they returned to Lake Louise. At the chalet, they waited until the storm subsided. Then the larger recovery party set out for the body of one of America's finest climbers.

Abbot was considered a cautious climber, and he climbed extensively in the Alps. The 30-year-old Boston lawyer was making his second attempt to climb the towering Mount Lefroy when he lost his life.

Last year, the unconquered mountain stumped Abbot, Fay and Thompson, all from the Appalachian Mountaineering Club in Philadelphia. They made two attempts using different routes over the course of two days in 1895. Problems loomed as a result of their late starts, the warm nights that left

the glacier snow soft and icy, the dripping meltwater and the dangerous, slippery conditions. On both occasions, they were defeated.

For their 1896 climb, Abbot and his group chose the Death Trap route. Allen had reached the col two years earlier from the Lake O'Hara side. The steep, narrow gully below it and between Lefroy and Victoria is prone to avalanches, and Allen chose Death Trap for the name of the gully and the col above it.

On the day of Abbot's fall, he and the other climbers awoke at 5:00 and were on the glacier at 7:30 a.m. By noon, they neared the col. Expecting an easy route from the col to the summit, they emptied their rucksacks of food.

In fact, they spent another 5 hours gaining altitude by zigzagging up ice slopes. Eventually, a rock buttress blocked their way about 75 feet from their goal. Given the late hour, Abbott decided to climb the vertical ahead of them rather than cutting more steps across the ice slope.

He had the others below him unrope so as not to endanger them with falling rock as he climbed. Earlier, the group's only other rope had been partly severed by sharp rock. Still, with the rope around him, Abbot climbed ahead and above the others. What happened is uncertain. Possibly, the limestone rock gave way under his feet. Whatever the cause, tragedy unfolded.

With the death of this famed climber, the Canadian Pacific Railway will undoubtedly pursue the idea of hiring Swiss mountain guides to accompany climbers and tourists staying at its mountain resorts. Already, some are suggesting that the ominous name Death Trap Col be changed to honor Phillip Abbot, who lost his life while attempting one of the world's great mountaineering challenges. ■

Photo

Up on the mountain during their mission to recover Abbot's body are (L-R) Tom Wilson, George Little, Willoughby Astley and Charles Fay. They reached the famed climber at about 4:00 p.m. today. He was partly covered by the terminus of an avalanche and by new snow.

Shoot-out Ends in Death

Two-year Manhunt Ends

MAY 29, 1897

Regina, SK — For the beleaguered NWMP in their standoff with a Cree man, Almighty Voice, help first arrived in the form of eight reinforcements. Today, another 26 Mounties were dispatched from Regina to the tragic scene. With them, the reinforcements took a nine-pounder gun. Other constables were also sent from Prince Albert. They took a seven-pounder field gun, and the fate of Almighty Voice and his two friends was sealed.

Today's final shoot-out was the last chapter in a NWMP manhunt that lasted almost two years. Both police officers and Natives are dead as a result of a tragedy that began with the killing of a cow.

When police recently tracked Almighty Voice and his two Native companions to the Batoche area, the fugitives escaped into the woods. Eventually, more than 100 police officers and volunteers surrounded them. During the fray, Constable Kerr, Corporal Hockins, postmaster/volunteer E. Grundy, Almighty Voice and his companions were all killed. Finally, after the two-day standoff and intermittent gunfire, police have blasted the area with field guns, and the fugitives are dead.

Two years ago, twenty-year-old Almighty Voice was arrested for stealing and killing a cow. He had applied to the Indian agent for beef to make broth for his sick child. The agent had refused. People on the reserve had little respect for him because he considered them to be troublemakers and he withheld rations as a means of control. He was also deemed to have acted unacceptably in relationships with Native women.

Although Natives were prohibited to kill Indian Department livestock without the agent's permission, Almighty Voice slaughtered the cow. Once arrested, he was taken to jail at Duck Lake. In all likelihood, his punishment would have been a few days in jail, but a guard told Almighty Voice in jest that he would hang.

His father was serving a six-month sentence for a lesser crime; his grandfather had died after seven months in jail. Rather than await his trial, Almighty Voice escaped.

Sergeant Colin Colebrook and a Metis tracker went in his pursuit. About 50 miles [80 km] east of Prince Albert, they caught up with the fugitive. Colebrook approached slowly and cautiously. Aggressive, angry and panicking,

Almighty Voice shot the officer. Because Colebrook died instantly, the tracker rode for help.

During the ensuing search, hostile Natives refused to assist authorities. Almighty Voice's mother hid him in an inaccessible dugout.

Eventually, dissatisfied with his hiding place, he became more daring. When he, his brother-in-law and his cousin killed another Indian Department steer, the police were again on his trail. There was a skirmish. The NWMP scout was wounded, and Almighty Voice became a public enemy in Canada.

Often warned when the police closed in on him, he escaped custody for nineteen months before the final tragic scene. Yesterday, two officers and 11 other Mounties trapped Almighty Voice and two friends at a heavily wooded area near Duck Lake. In the gunfire, both the inspector and the sergeant were wounded. Command fell to Corporal Hocking. While the Mounties combed the area, gunfire from the fugitives killed a postmaster/volunteer who was with them. More gunfire wounded the Corporal, who died this morning, and Constable Kerr, who died instantly.

Deep cultural misunderstandings and animosities may have given rise to the initial confrontation, and the loss of lives was a very high price to pay. ▪

Photo Left

Captain William Parker (standing) and Constable St. Dennis (seated right) were among the Mounties engaged in the final shootout.

Photo Right

Almighty Voice was the grandson of Chief One Arrow, who allied his band with Riel and the Metis during the 1885 rebellion. According to Natives of the One Arrow Reserve, the policies of the government, its Indian agent and NWMP punished them for that participation.

52 Dead!

Tragedy Knows No Borders

APRIL 10, 1898

Chilkoot Pass Summit, YK — Gold was the drawing card, but 52 people have died attempting to reach the summit of the Chilkoot Pass in the rush for Klondike gold. A week after the massive April 3 avalanche in the pass, all hope has vanished. A few still search for bodies, but rescue efforts have ended. It is expected that 63 will be confirmed dead after the spring melt.

The disaster took place just across the border, between Sheep Camp and the Scales. The tragedy is being felt by both countries. As some victims were Canadian, the North West Mounted Police was responsible for their bodies and the effects of the tragedy, including the task of notifying relatives. Many of the American victims had set foot on Canadian soil many times before.

Generally, gold seekers made the trip between Sheep Camp, Alaska and Canadian customs at the 3,000-foot [9,144 m] summit of the pass as many as 40 times. The Chilkoot Pass, more challenging than the nearby White Pass, is often considered the poor man's pass to the gold fields. To ensure the survival of gold rushers in the Yukon, Canada requires each person to bring a year's worth of supplies. Although there are two gasoline-powered cable lines for transporting supplies on the very last leg to the summit, prices are high. During the winter, those with little money haul supplies from sea level to the summit. By spring, they are ready to prospect. Carrying a 50-70-pound [23-32 kg] pack, most Klondikers can make one trip per day.

The slope begins at a 30-degree angle and steepens to 45 degrees. That pitch, plus heavy snow and melting, creates ideal avalanche conditions. The victims were just below the Golden Staircase. That 1,500-foot [457 m] ice staircase, created in winter from the feet of tens of thousands of people, is the last difficult climb to the Canadian customs house.

The day of the disaster, both the swirling snowstorm and the extreme avalanche danger made travel treacherous. The previous day, a small avalanche had broken away in the pass. At about 10:00 a.m. on April 3, just before the big slide, another small one broke off. Ten men were buried, but all were dug out. Despite that, 20 other construction workers were told to work on the tramway in the storm or collect their pay. Another 60 gold rushers at the Scales, a tent town halfway to the summit and just below the Golden Stairs, were unwilling to wait out the blizzard.

After weeks of storms, 6 feet [1.8 m] of wet snow covered the dry snow of early winter. Avalanche danger was so high that Native porters, who were usually for hire, refused to

work. Good weather and conditions were not expected. Four miles [6.4 km] below, Sheep Camp offered better weather and more amenities.

To avoid getting lost in the storm on their way down, the gold rushers, including two women, formed a life line. The line was a rope extended from the first to the last of them. Shortly before 11:00 a.m., having gone only a short distance, the doomed group was buried by the massive snow slide. The layer of wet, heavy snow ranged in depth from 5-30 feet [1.5-9 m]. Only a dozen people on the end of the rope were not swept away.

Despite the terrible conditions, survivors carried the news of the avalanche to the Scales, down to Sheep Camp and up to the NWMP. Rescuers arrived with shovels and sleds. Some victims who were not deeply buried had dug themselves out. One man was buried for 22 hours before he was found alive. A less fortunate woman who was discovered within 45 minutes was dead.

Time is all-important in avalanche rescue. Victims who live through the force and weight of the snow may die quickly of suffocation. Hours mean the difference between life and death. Despite rescue efforts eventually becoming futile, men continued the task of recovering fatalities for days.

Now, even the tramway power house at the Scales—the temporary way station for both the survivors and the dead—is emptying. Klondikers are once again carrying their packs up the Golden Staircase. The rush resumes, but the tragedy, rescues and recoveries in one of the world's most challenging landscapes will be remembered. ▪

Photo

For almost a week, rescue efforts and recovering bodies were the focus of all work. By April 6, more than 15 living victims were recovered from the snow, but only four survived. Here, the individual standing at the victim's head is believed to be a police officer.

Town Destroyed in Fire

B.C. History Wiped Out by Flames

SEPTEMBER 12, 1898

New Westminster, BC — Last night, a great tragedy befell the Royal City. Flames wiped out the business section, public buildings and residences of this historic city. The fire brigade, established in 1863 and currently including six full-time and six part-time firemen, made valiant efforts. However, the brigade was not adequately prepared to stop the firestorm.

Not kept in working order, the fireboat, intended for use in a waterfront disaster, proved to be useless. The department had a fire engine, but there were no horses available to pull it.

In some areas, firemen attached hoses to hydrants and turned them on. They were forced to flee, sometimes before turning them off, because of the aggressive spread of the fire. There was not enough water in the city's reservoirs.

Creating further havoc, both fire halls went up in flames. Personal efforts to save it were heroic. One fireman crawled into a burning building, and despite being blinded in one eye by flames, he carried a crying child to safety.

The Mayor called for assistance from Vancouver's fire department. That city sent 20 men, 2 hosereels and a horse-drawn engine so that water could be pumped onto the fire. Working together, firemen tore down small buildings in the path of the fire and finally, it was contained.

September 11 was to be the day for a celebratory fireworks display beyond anything that had ever been seen in the province. At 11:00 p.m. on Saturday, September 10, alarms rang to signal a fire. Starting from unknown causes in a hay-filled warehouse on the waterfront, flames engulfed commercial and government buildings, docks and homes.

Two warehouses, one for coal and one for livestock feed, created the intense heat that complicated firefighting. Wooden sidewalks became avenues of flame. Worst of all, along the wharves, three steamers caught fire. When the ropes of their moorings burned, the steamers became floating bonfires and carried the fire to still other locations. It reached its peak at about 3:00 in the morning.

To help the homeless, private and military donations of about 1,500 blankets and 180 tents are on their way. Other help is expected soon.

Unfortunately, little could be saved and the

flames devoured much of B.C.'s earliest history. Vancouver was a very new community with small buildings when it burned in 1886. As a result, it could be rebuilt quite easily.

New Westminster, named by Queen Victoria, was the first city to be incorporated in the province. The Royal Engineers, who arrived in 1855, built many of the early buildings that were destroyed. New Westminster was the mainland capital city from 1859-1868. Only when the crown colony of British Columbia united with colony of Vancouver Island did Victoria become the capital, and the fire has claimed much of this history.

About 300 buildings were lost. Of the early colonial buildings, only one of the elegant homes was spared. Built in 1862 and belonging to an early steamboat operator, the two-storey, fourteen-room home contains heirlooms such as an autographed engraving of Sir John Franklin. The rest of New Westminster's history has gone up in flames, including treasured artifacts pertaining to John Franklin and Captain Vancouver, which were at the library.

The question now is whether New Westminster has a future. The fire has dealt a devastating blow. Vancouver helped in a time of need, but now the sister city and competitor is willing and able to provide services to the Fraser Valley and the B.C. coast. ▪

Photo

A seven-block area was destroyed in New Westminster. Warehouses, two canneries, the CPR station, city hall, post office, court house, public library, most churches, all hotels, the newspaper building, two banks and the opera house are among the public and commercial buildings that were destroyed.

Fire Strikes Again

Will Dawson Rise from Ashes?

JANUARY 11, 1900

Dawson, YK — This time, the fire brigade was well organized to fight the latest fire raging through Dawson. Although it was not nearly as devastating as last year's fire, important buildings were razed.

Once Canada's largest city north of Vancouver and west of Winnipeg, Dawson is less than 200 miles [322 km] south of the Arctic Circle. Despite temperatures that plummet to minus -80 °F [-64 °C], the lure of gold has kept the community vibrant. With its history of fires, Dawson knows how to rebuild quickly. Last year, after its worst fire, residents rebuilt half the town.

On the evening of April 26, 1899, the temperature was minus 45 °F [-45 °C]. The fire broke out when a dance-hall girl dropped a lamp. It was 7:30 in the evening, just when entertainment at the saloon was in full swing. As flames began to rage through wooden buildings, people ran naked into the streets.

Firemen were on strike for higher wages. There was no fire in the boiler of the steam engine, and firefighting efforts were in complete disarray. A bucket brigade and hand pumps were the only option. With the freezing temperatures and high winds, conditions were almost intolerable. In fact, it was so cold that men wearing parkas and furs did not feel the heat when fighting the inferno.

The men created a water source by building a fire on the river to thaw a hole in the ice. To halt the spread of the fire, they blew up buildings in its path. While some were trying to save the town, others were more interested in saving the buildings they personally owned. Businessmen offered thousands of dollars to anyone who would help them.

The Fairview Hotel was one of the very few lucky buildings. Wet blankets and bedding were hung from the roof and somehow, that saved it. Ironically, ice blocks already stored for the next summer's use under sawdust insulation also remained solid and untouched.

Eventually, the fire spread from the river to the swamp, where it burned itself out. One hundred eleven buildings were destroyed in the town of 30,000 people. Damages were estimated at about $1 million, and for rebuilding, merchants charged 25 cents per nail.

Still another fire broke out in October of 1898, just after the town had purchased its first steam engine for firefighting. Due to the town's repeat-

ing history of fires, the community now keeps its fire department in a state of constant readiness.

Unfortunately, residents must be prepared to rebuilt. As many of the miners and prospectors have moved to the gold fields in Alaska, they have taken with them their money and their optimism. Thus, the northern community faces a challenge. However, the residents have proven their perseverance, and Dawson is likely to rise again. ▪

Photo

As a result of the January 10, 1900 fire, the Canadian Bank of Commerce is in ruins. Outside, a police officer stands guard to prevent looting. When the Bank of British North America burned during the April 26, 1899 fire, the vault melted, exposing gold dust and nuggets in storage. The heat was so intense that the gold also melted. By midnight, seven NWMP were standing guard to prevent theft. Because of the strong presence provided by the NWMP under Commander Sam Steele and his predecessor, C. P. Constantine, there was little threat of serious lawlessness in Dawson.

Mountain Collapses!

Miners Buried Alive

APRIL 30, 1903

Frank, AB — Today, there are many stories of heroism at Frank, a coal mining community near the B.C. border. At 4:00 a.m., massive boulders of limestone broke away from Turtle Mountain. The April 29 slide took less than 90 seconds.

Miners had the courage to save themselves. They were trapped underground after rock buried the mine and the community. Unbelievably, they put aside their panic and dug their way to safety.

The slide sealed the entrance to the Canadian American Coal and Coke Company mines, and the weight of the rocks collapsed the tunnels. Yet deep underground, in the darkness and despite their fears, these courageous men worked under unthinkable conditions with little air and water and no food. At about 5:00 p.m. on April 29, after almost 13 hours of being buried alive, seventeen miners walked away from almost certain death.

When the slide occurred, the wind blast was powerful, even underground. Some miners were blown against tunnel walls. Others were injured by crumbling or falling rock. Aware of the threats of another explosion and the accumulation of carbon dioxide—the deadly gas referred to as "afterdamp"—the men rushed to the mine entrance.

They were entombed by twisted timbers and rock. Desperately seeking escape, they checked the entrance of the lower air-supply tunnel. That avenue of escape had been flooded. Another airway also proved to be blocked, and coal gas was accumulating. After three hours, most realized they had no way out. With their air-supply shafts sealed, death was close at hand.

Heroically, the men used the strength they might have expended in panic and directed it towards survival.

Above ground, rescuers were likely trying to dig them out of the prison. However, those buried knew how little time they had, and they knew how entirely the mine entrance was blocked. Left to their own devices, the imprisoned miners got organized to save themselves.

By 7:00 a.m., the men selected their site and began digging. Joe Chapman, Dan Mackenzie and Charlie Farell are all credited with leadership during the dark time of fear, exhaustion and uncertainty. Although it seemed futile, the arduous task would only be successful if they worked together. The strength and persistence of all were needed.

The digging continued all

day in the cramped shaft. Two or three worked the escape tunnel. Once they were exhausted, they were replaced by other men. They were uncertain as to how far they would have to dig and exactly where they would surface. The men dug through 20 feet [6 m] of coal and another nine feet [2.7 m] of limestone rock. At 5:00, they were free. The buried miners surfaced just uphill from the rescuers.

Immediately after the rumbling that woke up residents of the area, rescue efforts became intensive. Realizing that as many as 20 men were working the night shift at the mine, other miners and townspeople searched rubble for the entrance to the workings. Once they finally found it, they began digging towards the underground victims. Almost 13 hours had passed since the slide, and those involved in the rescue effort had not moved much rock. For everyone, there was unbelievable relief when the buried miners surfaced.

But their joy was short-lived. Elsewhere, others are still missing. Tons of rock cover more than a square mile [3 km^2] in the valley. Averaging 45 feet [13.7 m] in depth, the slide crushed seven homes and other buildings on the edge of town. Two ranches, a construction camp and a railway camp have been devastated.

Today, search and rescue continues, but the slide is believed to have claimed the lives of 76 people. The entire crew of a construction camp is presumed dead. Twelve men at a railway camp have died. Two miners and a weigh man, who were above ground during the shift, are missing and presumed dead.

For many, elation is mixed with sorrow. One man who escaped the mine has discovered that his home was crushed by rock. In it were his wife and seven children. However, others have been rescued. One baby girl and her two sisters were found alive. Unfortunately, their parents and four brothers did not survive.

Fortuitously, 128 men who were scheduled to build a railway spur line at Frank during the time of the slide did not arrive. A mix-up delayed them and saved their lives. Today, those men, 25 NWMP and two doctors and nurses from Fort Macleod joined in the endless effort to alleviate the effects of the disaster. ▪

Photo

Due to the slide, telegraph lines to the east are down. Early this morning, the unsuspecting Spokane Flyer thundered towards tragedy. In the dark, Sid Choquette struggled across the endless giant boulders. By flagging down the train, he saved passengers and crew.

Murderer Finally Hangs

Justice Has Been Done

FEBRUARY 2, 1904

Calgary, AB — Nine days after his January 24, 1904 arrest, murderer and escaped convict Ernest Cashel was hanged. The Wyoming-born criminal murdered one man and terrorized the central and southern Alberta territory for a year and a half. His arrest involved 16 NWMP officers, 6 soldiers from the Canadian Mounted Rifles and 16 citizen volunteers who bravely faced the desperado's gunfire.

His life of crime in the territories began with bad cheques. He was arrested near Ponoka, but in October of 1902 he escaped custody. While being transported by train to jail, he went to the washroom on the train. Minutes later, Calgary's police chief found the window open and Cashel gone.

By late November that year, he became the suspect in a much worse crime. An elderly homesteader in the Lacombe area was missing. He was presumed dead, but there was no body. NWMP Constable Alick Pennycuick became the celebrated investigator for the murder case. He suspected Cashel. He learned that the victim had sheltered a man whose alias was one that Cashel had previously used. A gold certificate and other property were missing. In the same area, other ranchers reported missing property. One rancher loaned a young man a horse and saddle, supposedly to use while looking for his own pony, which had strayed. The horse was never returned. All reports about the slim-looking young man of about 20, who Pennycuick believed to be Cashel, became useful clues.

In the meantime, Cashel was continuing his life of fraud and property crime. Finally, after three months on the loose, NWMP captured him at Anthracite near Banff. Cashel was found guilty of stealing a horse and a diamond ring. Although Cashel was suspected of the Lacombe rancher's murder, the victim's body was still missing. Without the murder charge, he was sentenced to only three years at Stone Mountain Penitentiary in Manitoba.

Finally, the body of the homesteader was discovered. In July of 1903, the murder victim washed up on the shore of the Red Deer River. Pennycuick matched the bullet in the body with Cashel's gun. The desperado was charged and returned to Calgary for trial. The trial lasted nine days, but the jury took only 35 minutes to pass their

guilty verdict. He was sentenced to hang on December 26.

While waiting in the Calgary jail, Cashel's brother smuggled two revolvers to him. On December 10, 1903, when he was taken from his cell so it could be searched, Cashel threatened the three officers with the revolvers. Locking them in the cell, he unlocked his shackles and escaped.

His final confrontation with the police took place at a ranch about seven miles [11.2 km] west of Calgary. Although he might have escaped Canadian justice by fleeing to his homeland, he stayed in the Calgary area. He was determined to free his brother, John, who was convicted of aiding him in the escape.

The manhunt that followed was unprecedented in the Alberta territory. On the day of his final arrest, Cashel was discovered hiding beneath a trap door in a bunkhouse. An unarmed Mountie attempted to negotiate a surrender, but Cashel responded with two shots. Once armed, the officer returned fire, shooting Cashel in the heel while the fugitive stood on the staircase.

Once again, Cashel fled to the cellar. When he shouted threats of suicide, Inspector Duffus ordered the bunkhouse burned. With his hiding place in flames, Cashel surrendered.

He was returned to jail in Calgary, and as his hanging is today, fearful people in the community believe that justice has been done. ■

Photo

When Cashel escaped from the Calgary jail after being sentenced to hang, a reward of $1,000 was offered. This time, he eluded custody for 45 days, but the Mounties finally got their man.

Rocky Mountain Resort in Ruins

Employees Jump to Safety

JULY 3, 1904

Lake Louise, AB — Three employees, two Japanese and one Caucasian, jumped from a second-storey window of the Chateau Lake Louise. One cut an artery in his hand, one suffered burns to his hand and the other has a broken leg.

On the north wing roof, another employee found himself in trouble. He had climbed out a window and onto the very steep roof, only to find that he was surrounded by fire. When he found that he could not return, it looked like certain death. Capable and quick-thinking, he used the axe he had carried from the building to cut foot and hand holds in the roof and made his way to safety.

Some of the guests, all of whom escaped safely, sat in chairs that had been removed from the hotel and watched both the careful movements of the man on the roof and the conflagration. A few lost everything except their clothes, but most did not suffer serious losses. The steel vaults of the old wing remain red hot, but management expects that the tickets, money and valuables of the guests will not have been damaged.

Some guests helped fight the fire and remove property from the hotel. They and employees threw things from the windows or carried out items as they escaped. With the exception of the piano, most of the items that were saved were inexpensive.

No one was seriously hurt, and fire-fighting efforts became the priority. With the telephone exchange destroyed in the burning building, there was no way to get immediate help. Only by sending a driver to the CPR depot and telegraphing a message to the outside were the threatened able to get help.

Within an hour of learning of the disaster, the hotel's construction superintendent drove from Banff to Lake Louise and was on the roof fighting the fire. The old wing of the resort was in flames, but the concrete wall of the newer part of the hotel acted as a fire barrier. However, there was still a risk of sparks hitting the roof and igniting that section.

The other great danger was that the fire would spread into the surrounding forests. Fire Chief Hill and game wardens hurried to the scene. Fortunately, with ample water, guests, employees and fire fighters were able to contain the fire to the hotel and a

few trees.

Dry, hot weather and electrical storms have meant that forest fire danger is very high. In Alberta, forest fire danger is high in the Bow reserve, the Crowsnest and the Brazeau. Smoke has drifted into the province from fires in British Columbia.

Near Sicamous, there are three large and seventeen small fires burning. Near Revelstoke, one large and five small fires are burning. There are seven small fires in the Kamloops area. Many small ones are burning on Seymour Arm. Other areas such as Golden have assigned extra fire patrols. Most of the forest fires were caused by lightning. Since many of them occur in rough or remote areas, fighting them is difficult. ▪

Photo

At the time of the fire, there were 300-400 registered guests in the new section, which was not damaged. The 40-50 guests in the old wing were provided with accommodation at the Banff Springs Hotel.

Heroine Saves 38 Lives

Mother of Five Ignores Risks Again

DECEMBER, 1906

Cape Beale, BC — 38 crewmen of the American barque *Coloma* were saved from drowning because of the bravery of Minnie Patterson. For the second time in little more than a year, Patterson has saved the victims of shipwrecks. This time, she braved a west coast gale to get help for the floundering *Coloma*.

Telephone lines were down in this isolated coastal area when the lighthouse keeper realized that the *Coloma* was in serious trouble. It had drifted near a reef at Cape Beale. Despite the rain and fog, he could see that the fore and mizzen masts had been wrecked and that the ensign was being flown upside down, a signal of distress. With rough seas tearing at the vessel, the crew was dependent on the main mast for survival.

The ship needed help, and there was no method of communicating the impending disaster. The Canadian government's *Quadra*, a lighthouse tender capable of accomplishing a rescue, was at Bamfield.

Leaving her husband, who had to remain at the foghorn on duty, Minnie set out on foot for help. It was night, and in the rain and darkness, she carried a lantern as she defied the hazards. Uprooted trees blocked the rain-drenched trails and the slippery, protruding rock on the trail was perilous.

If she had been hurt on the journey, there was only her dog to help her. But the woman who had recently given birth to her fifth child covered almost 10 miles [16 km] to save the endangered crewmen. The nearest source of help was James McKay at Bamfield Creek. Here, too, phone lines were down, and Mr. McKay was out repairing them.

Minnie's friend, Mrs. Annie McKay, was the daughter of a lighthouse keeper and understood the urgency of the situation. The two women stepped into a small boat, picked up the oars and rowed out to the *Quadra*. Once the message of distress was conveyed, the *Quadra* set out in rough seas. When the *Coloma* was sighted, the second officer took a small boat to the threatened vessel and the crew was rescued. Just after the Americans were safe on the deck of the Canadian steamship, their

barque wrecked on rocks.

Minnie Patterson's dedication and courage first became evident to others very early in 1906. In the bitter cold of January, the S.S. *Valencia* hit a reef near the Cape Beale lighthouse. At least 117 people drowned in the disaster. Survivors were brought to the Cape Beale lighthouse and Minnie's home. She was pregnant and had to care for her own children. Still, for 70 hours, never stopping for sleep, she nursed and cared for the needs of the survivors. ▪

Photo

In bad winter weather, the rough sea and rocky coastline near Cape Beale become a hazard for ships. Here, the surviving crew of the tragic S.S. Valencia, *which was wrecked on January 23, are brought ashore from the HMS* Eqeria.

Sternwheeler Destroyed

Rivers Wreak Havoc

JUNE 7, 1908

Saskatoon, SK — Twelve people escaped death when high water on the South Saskatchewan River claimed the sternwheeler named *The City of Medicine Hat*. At noon today, engineer Mike McKeon was forced to dive from a lower deck of the ship. He hoped to reach the safety of Saskatoon's traffic bridge, but he was carried about half a mile [0.8 km] downstream. When rescuers reached him, he was in a state of exhaustion.

All other passengers and crew were on the upper deck. They leaped to the bridge without injury. No sooner were they off the vessel than the *Medicine Hat* keeled over into the river.

These circumstances evolved as a result of this year's flooding river. Currently, the water level is within one foot [0.3 m] of the high-water mark for the great flood of 1902. That level is six feet [1.8 m] above the high-water mark for last year.

The doomed steamer left Medicine Hat for Winnipeg on Saturday. The owner, Captain Ross, set sail with a crew of seven in addition to himself. Only four were aboard for the pleasure cruise. After traveling a record 225 miles [362 km] in 24 hours, the sternwheeler settled to a leisurely pace as it continued to Saskatoon.

When the captain reached the Grand Trunk Pacific Bridge on the outskirts of Saskatoon, he moored his steamer. It was riding so high that the smokestack would not pass under the bridge. Late in the day, the crew decided to dismantle the stack. Otherwise, the rapidly rising water might have kept them there for some days.

With the smokestack no longer presenting an obstacle, *The City of Medicine Hat* passed under the Grand Trunk Bridge. At the Canadian National Railway Bridge, the captain made careful measurements. Once again, his vessel passed safely.

Next, it reached the new traffic bridge. There, the steamer hit a loose telephone wire that stretched across the river. The wire snapped and coiled around the pilot house. It toppled, and the captain narrowly missed being caught under its timbers.

The crippled ship was swept by the rushing water against the steel girders of the bridge. All but one of the crew and passengers jumped to safety.

In the meantime, the coal barge towed by the sternwheeler had broken away. Carried by the swift

current and high waters, it capsized downstream, dumping tons of coal in the river.

The South Saskatchewan River has caused problems not only for *The City of Medicine Hat,* but also for the town that is its namesake. There, some have had to abandon their homes. One of the most distressing sights was that of a two-storey house being carried down river by the floodwaters.

The Oldman River flows into the South Saskatchewan just west of Medicine Hat. As both the Oldman and Belly Rivers are also flooding, Lethbridge is suffering. High water washed out one of the small bridges near town, the approach to the Big Belly bridge and a CPR pumping station. The CPR trestles at Whoop-Up and St. Mary's are endangered, and train schedules have been cancelled.

Pincher Creek is flooded. The traffic bridge at Cardston has been damaged and the railway bridge washed away. At Fort Macleod, the Oldman River has carried away the approach to the C & E Railway, and many buildings are under water. Flooding has undermined the foundation of the town hall and threatens the hospital and telegraph lines.

Throughout the drainage basin of the South Saskatchewan, floods are wreaking havoc, but so far, no lives have been lost. ▪

Top Photo

The City of Medicine Hat, *built in 1906 for Captain Ross and specifically designed for river navigation, cost about $35,0000. The steamboat was licensed to carry as many as 250 passengers.*

Bottom Photo

Given the sternwheeller's damage and its position in the river, it seems highly unlikely that she can be righted and restored to a seaworthy state. She was insured only for fire. Her owner, Captain Ross, a former captain in the German navy and a capable navigator, has had little luck with his sternwheelers. This year, his other steamboat, the Assiniboia, *was crushed by lake ice.*

Train Crash Kills Ten

Phantom Train Foretells Death

JULY 9, 1908

Medicine Hat, AB — When two locomotives collided in a head-on accident near here today, the fireman of the single engine jumped to safety. The wreck happened close to the CPR railway yards, where he ran for help. Workers immediately rushed to the scene and began the difficult task of rescuing victims in the tangled mass of steel. Still, the accident has cost lives.

In the crash, three trainmen and two passengers died instantly. Although they were miraculously pulled from the wreckage and rushed to hospital, many other passengers were unable to recover. Five other passengers have died. Many others were injured and some remain in serious condition in hospital.

To date, the train wreck is the worst in Alberta's homestead history. It was about 8:20 a.m. when the two engines collided. Twenty minutes earlier, a CPR single engine had left Medicine Hat on a scheduled run. It was to couple with the Spokane Flyer at Coleridge and take the Flyer to Moose Jaw, Saskatchewan.

At a sharp curve on a downgrade within a mile [1.6 km] of the Medicine Hat train yards, a passenger train collided with the single engine. Train 514 was en route from the Crowsnest Pass through Lethbridge, and the passenger train was running two hours behind schedule.

In the collision, both engineers were killed. On the Crowsnest Pass train, the fireman, conductor and seven of the passengers lost their lives. The passenger fatalities were riding in the tourist or colonist car. Those in the first class car, which did not leave the track, were simply shaken up.

A homesteader witnessed the tragedy. From where he stood, he saw the two trains on their collision course. Because of the high-cut bank and the curve, visibility was blocked for the engineers of the two approaching trains. The farmer waved his arms to warn one of the engineers of the impending disaster. Whether the signal was initially misunderstood is uncertain. By the time the engineer reversed his engine, it was too late to prevent the collision.

Apparently, before leaving the station, the engineer on the single engine should have been told that the scheduled arrival of the passenger train had not been recorded. However, receiving his clearance papers from the operator

on duty, he proceeded from the station.

The fireman who survived said that he saw the headlight of the approaching train. The light was so bright that the firebox of the locomotive appeared to be open. By the time he noticed the other train, disaster could not be averted.

As if they were being warned of their fate, two of the trainmen witnessed a ghostly foreshadowing of the collision. Days before, while on scheduled runs, both claimed to have seen a passenger train approaching their train at the very point of the accident. However, just before they collided with the phantom train, it veered off on nonexistent tracks. The ghostly train carried passengers who waved to them.

Not believing their own eyes, the two admitted the vision only to each other. The engineer who died on the single engine was one of these two men. The other was working in the CPR yards at the time of today's tragedy. ■

Photo

In the tragic collision, the light engine was derailed. The engine of the passenger train, which rolled off the track and into the ditch, is a write-off. The boiler head of the other engine was torn off. The mail car, baggage car and a tourist car on the passenger train have crumpled into a space that would normally fit only one railcar.

Fernie Destroyed in Fire

Trains Become Lifeline

AUGUST 2, 1908

Fernie, BC — A town of close to 6,000 has virtually disappeared. The Crowsnest Pass area has faced another disaster, but pledges of help are coming from cities and towns across Canada as well as from Americans.

Yesterday, despite acts of courage during the Fernie inferno, ten lives were lost in the area from Fernie to Hosmer. This time, the culprit was an unattended slash-pile fire. At Cedar Valley Lumber Company, it had slowly burned for the past month. Suddenly, it leaped into flame. By 3:00 p.m., it had destroyed the mill. Then the wind picked up to an almost cyclonic velocity. The flames became a firestorm and spread through the forest to the Elk Lumber Company yard at the west end of Fernie.

For three hours, panic and courage raced side by side through the streets. The fire department's attempt to control it was futile. The conflagration spread to the wooden buildings of Old Town and then the New Town areas of Fernie.

Trains became the only means of escape. Fleeing to the Great Northern Railway station, about 1,500 people boarded as many as 30 box cars, flat cars, cattle cars and coal cars to be whisked away to safety along the river. Eventually, some were moved as far away as Michel.

In town, the Western Canada Warehouse survived the firestorm, and at times it was refuge to nearly 100 people. There, men on the roof kept soaking the shingles, and the people they protected were spared. Many were patients from the general and isolation hospitals. Given top priority, nurses and other volunteers had moved the patients to the concrete wholesale warehouse. As the flames approached this refuge, the patients were again moved to the railway tracks. Their final rescue took place at midnight when a railcar arrived to take them to Cranbrook. A good neighbour, that community offered refuge not only to hospital patients but to many victims of the tragedy.

Some people found safety in the stone office of the Crow's Nest Pass Coal Company building. Although they survived, the sweltering heat, the black and smoke-filled sky in midafternoon and the flying ash were ominous threats throughout the ordeal.

The CPR and Great Northern tracks were eventually twisted by the infer-

no. The CPR lost all facilities, track and cars in the area. The station and water tank of the GN were saved, but coal cars at its depot melted in the fire. As soon as those tracks could be made passable, relief was rushed to survivors.

Within 40 minutes of a plea for help for the homeless, Cranbrook had a trainload of tents and supplies on its way to the disaster area. Once relief tents were erected at one of the coal company yards, victims were provided with free food, clothing and shelter. Since Fernie is the next-door neighbour of southern Alberta's coal mining towns, Albertans are already offering their aid.

For Fernie, this was the third fire in four years. In the spring of 1904, a fire started in the early morning at the general store. In 1905, another started in a tailor's shop. Still another, a month later, caused $40,000 damage.

Coal mining disasters at or near Fernie have had grave consequences, as many lives have been lost. On the day previous to the great fire, an explosion at the Coal Creek mine trapped 23 miners. Three died immediately. It was eight hours before the others could be rescued, and consequently, one miner later died of injuries. The rest faced the firestorm that has devastated the community.

Despite Fernie's many sorrows, it has demonstrated its perseverance. Like other survivors with heart, the town will rise phoenix-like and start again—with the help of its friends. ■

Photo

This church was only one of about 1,000 wooden buildings razed by the fire. The only buildings to remain standing were 23 homes, the office of the Crow's Nest Pass Coal Company and the Western Canada Warehouse.

Avalanche Buries CPR Crew

Rescuers Search for Survivors

MARCH 7, 1910

Rogers Pass, BC — Rescue work continues after a second snow slide tore down Mount Avalanche at 11:30 p.m. on Friday, March 4. It claimed more lives than any avalanche disaster in Canada. For three days, Canadian Pacific Railway crews and volunteers from Revelstoke have searched for bodies and survivors, but the death toll is now estimated to have reached 62. Victims include both white and Japanese CPR crew members who were clearing track after the first of the two slides. Not all the bodies have been found, but those still missing are assumed to be dead.

News of the big slide reached Revelstoke at around midnight, shortly after the avalanche. Once the town's fire bell and the steam whistles of the CPR roundhouse announced the S.O.S., about 150 railway workers and volunteers boarded a rescue train.

Unfortunately, the avalanche killed almost everyone working at the site of the slide. CPR road master John Anderson, who was an eye witness to the slide, claims that although it was almost midnight and storming, crews were still clearing the track of a smaller snow slide. At the 4,300-foot [1,311 m] level of the pass, seven feet [2.1 m] of snow fell in nine days. Because that area accumulates an average annual snowfall of 33 feet [10 m], there was about 30 feet [9.1 m] of snow even prior to the snowfall. Avalanche danger was extreme, but there had not been a slide off Mount Avalanche in 11 years. When a small slide blocked the track, crews were sent to clear it.

As the men worked, lookouts were on duty, but the blizzard meant that visibility was negligible. Anderson left the site and went to the watchman's cabin to dispatch a message that the job would be finished in about two hours. He was just returning when, above the noise of the storm and rotary snow plows, he heard the roar of the wind. Finally, realizing that a massive slide had poured down Mount Avalanche and buried the crews, locomotive, rotary plow and the other equipment, he floundered though the snow.

He heard a cry for help and made his way over the chunks of ice and snow. On top of a snow shed was E. LeChance, hurled 60 feet [18.2 m] from where he had been firing the locomotive of the rotary plow. He was half buried, but Anderson managed to dig him out and then went

back to the watchman's shed to phone for help.

Bridge carpenter D. McCrae, who was swept away by the avalanche, found that he was able to dig himself out.

In rescuing survivors of an avalanche, time is always of the essence, and victims must be found before they suffocate in the snow. McCrae assisted in the rescue of the injured LeChance and then began to search for others. Although he found W. Phillips, the engineer was pinned under the locomotive and could not be saved.

Anderson found a survivor from one of the Japanese work crews. The cook was disoriented but uninjured. Also safe were two linemen who had been working in the area but had just left the slide site. A Japanese crew of 10 had also left the slide path to go for supper.

Those at the site and arriving from Rogers Pass Station did as much as they could to find and dig out victims, but in the darkness and the fierce storm, the task was monumental.

During the first day, the snow had to be cleared with shovels because there were bodies on the track. Only after that could the rotary plow be used to clear the track for relief and supply trains. ▪

Photo

Although they are well trained and experienced at digging out avalanche areas, the CPR's rescue crews had never dealt with a catastrophe of such magnitude. Whether or not they were paid for their efforts, railway workers and residents helped with avalanche rescues out of their sense of honour and duty.

Derailed Train Crashes to Ice

Leaking Fuel Feeds Fire

MARCH 5, 1912

Saskatoon, SK — Yesterday, the efforts of firemen, Mounties and volunteers saved lives. At 6:00 p.m. on March 4, a sleeping car from Regina's Capital City Express containing 13 people crashed through the Canadian Northern Railway Bridge to the ice. The car, called the Kipling, derailed, took out a span of the bridge and fell to the ice as it crossed the South Saskatchewan River.

The sleeping car landed on its side in a tangled mass. During the accident, the acetylene tank that fuelled lighting for the car was damaged. Leaking, it burst into flame. Quickly, the wood structures and fittings in the car created a perilous blaze.

Train crewmen, Mounties, city workers and volunteers jumped immediately to the rescue. They knocked out windows in the car and one by one, victims were lifted from the wreckage. One of the men who went down with the car and who suffered only slight injuries helped other victims out of the wreckage, but as the fire became more threatening, he left the car.

When firemen arrived at the scene, all known victims had already been pulled from the burning car. However, their uncertainty as to how many passengers might remain in the berths led firemen to drop though an open window to look for other victims.

Once all the survivors were out of the car, firemen faced the difficult task of putting out the blaze. Due to the increasing likelihood that the tank would explode, the work was life-threatening. Firemen used four hoses to fight the blaze. In order to extinguish it completely, they had to focus on the fuel source of the leaking tank. Men manoeuvred in order to approach the tank from the rear of the car, draping it with soaked clothing and blankets from inside the car. With the leakage contained, they were able to tackle the fire itself.

While battling the flames, additional dangers threatened the firemen as they worked. Timbers still supporting the bridge had been destabilized. Had the bridge or fallen timber which was propped precariously around the wreckage collapsed, the men might have been crushed. By 7:10, the fire was out.

Twelve of the those rescued from the Kipling will recover. Their injuries range from minor cuts and bruises to rib fractures. One man has lost his life as a result of a fractured skull. He was a traveling

salesman for Fry's Cocoa and is the father of a child born only a few days ago.

Most of the Kipling's passengers were traveling salesmen or businessmen. Three Anglican ministers were also aboard, one of whom is a railway missionary who travels to communities that lack a resident minister. His injuries were minor and he expects to return to work within days. The only woman aboard was uninjured.

Had the disaster happened after the ice melted, none would have survived, given the depth and current of the South Saskatchewan River.

The accident happened as the train moved out of the railway yard and onto the railway bridge. The last few cars were derailed. Pulled onto the bridge by the locomotive, the derailed cars knocked out one of the bridge's twelve-inch [30 cm] supports. Weakened, the span gave way as the Kipling, the last of the cars, crossed the bridge. ▪

Photo

A steam crane will remove the wreckage and repair the railway bridge after the Kipling disaster.

Railways Forced to Obey New Rules

Toughest Regulations on Continent

MAY 22, 1912

Victoria, BC — The British Columbia government has won a victory for its forests and dryland interior. Beginning today, any fire that starts within 300 feet [91 m] of the railway track is presumed to have been caused by the railway. Locomotives can no longer ignite forests and prairies without the railway companies' accepting responsibility.

The strongest lobby for revised regulations in the Railway Act, B.C. has finally convinced Ottawa to respond. The revised act will also appease settlers. With these revisions, Canada boasts the most forceful regulations pertaining to railway fires in North America.

Since it rains very little on much of the prairies and in the interior of B.C., fire is a major hazard. Settlers have long blamed coal-burning locomotives for the devastating fires that sweep through Western Canada. Newer communities have only rudimentary fire-fighting equipment, and few have paid fire fighters.

Ongoing revisions to the Railway Act of 1903 have demanded action on the part of railway companies. The newest rules require companies to hire fire rangers to patrol along the lines. Telegraph operators must record each time a velocipede passes on fire patrol. These patrollers must carry two shovels, two canvas buckets and an axe. Foot patrollers are required to carry shovels and canvas buckets. As well, an axe, four buckets and three mattocks must be stored at the tool shed of each section. All costs for equipment are to be borne by the railway.

In addition, rules regarding fire-prevention appliances on locomotives are more rigorous. Railways are to ensure that they have cleared rights of way and that all live coals emptied on those rights of way are entirely extinguished. They are to build fire guards on their own land, government land and individually owned lands along the rails. Railways must also reimburse finan-

cially anyone who is subject to fire losses resulting from the railway.

Although B.C. has been the strongest proponent of the revisions, the amendments are consistent with public opinion throughout the West. When winds combine with drought conditions, prairie fires sweep across immense stretches of land. While many fires are inadvertently caused by lightning, mishap or carelessness, settlers feel that the railway fires can be prevented.

As far back as 1894, the people of Moose Jaw circulated a petition demanding the changes that are now part of the 1912 regulations. Two years earlier, in April of 1892, an Alberta prairie fire raged from Gleichen, near Calgary, as far north as Red Deer. That central Alberta community was barely saved from destruction, and every male member of the community was needed to fight the fire. Such uncontrollable walls of flame are among westerners' greatest fears, and the strict new regulations are welcomed. ▪

Photo

Requiring railways to provide fire patrollers was in 1911 considered to be progressive, even radical. The new 1912 rules are more demanding and specific. They regulate the number of patrols, foot patrolmen and velocipedes for each subdivision from Manitoba to British Columbia. The rules even specify which railway lines patrols must follow 30 minutes after a train's passing.

Tornado Shatters Regina

28 Dead, 2,500 Homeless

JULY 1, 1912

Regina, SK — Today, the volunteer militia of the 95 Rifles, 16 Mounted Rifles and 26 Rifles arrived to help residents of Regina recover. The militia volunteers were performing summer exercises nearby when yesterday's tornado dealt the community a crippling blow. The volunteers will assist the homeless and sift through the debris in search of anyone who may still be buried under the rubble.

On June 30 at 4:50 p.m., the violent whirlwind flattened buildings in a five-block area of the city's downtown. The area included the business district, high-end housing and the CPR yards. At least 28 are dead and an estimated 200 were seriously injured. The homeless number about 2,500.

Winds at the vortex are estimated to have moved at 100-500 miles [160-805 km] per hour. The force of the whirlwind destroyed about 500 buildings, creating massive piles of rubble. Residents of this prairie community have responded to the emergency by rescuing tirelessly their neighbours and strangers alike.

Before the disaster struck, the day was oppressively hot. By afternoon, a light rain was falling. The dark funnel cloud was first noticed moving southwest to northwest through the area. Then, the cyclone began to move in a more northerly direction, hitting downtown Regina.

News of the disaster was communicated to RCMP at the barracks on the edge of town. Immediately, the force mobilized to help. One hundred fifty Mounties and recruits moved into the area to assist rescue efforts and to help victims.

The tornado was so devastating that the community needed more help. With telegraph and telephone lines down, local people saved lives while the lines were being restrung.

One rescue effort was directed toward those trapped in a large rooming house, despite which the fatalities in the area were significant. Eight people were in the house; three died and five others were injured. Another man died outside the boarding house.

John Gardner worked to save a family that had taken refuge in a cellar. Of the nine victims trapped there, eight were rescued and suffered only minor injuries. The family's son died from fallen debris.

In another rescue attempt, Alex Huston of the Municipal Department organized passers-by to try

to free a trapped Chinese man. However, the elderly gentleman died before they could reach him.

People used whatever vehicles were available as improvised ambulances, and the General and Grey Nuns Hospitals filled with people who had been pulled from the wreckage. Additional emergency medical attention was available at a parish hall, Immigration Hall, CNR freight sheds and two business buildings.

The first assistance came at 7:15 p.m. yesterday. The speedy efforts of Regina's neighbour, Moose Jaw, brought doctors, nurses and medical supplies by train. Today, additional medical help from Winnipeg is expected.

Electrical workers are among the unsung heros. Because the tornado had brought down electrical wires, live wires created a serious hazard. As evening approached, a demoralized city and exhausted medical and rescue workers faced darkness.

Unbelievably, the lights came on by 9:00 p.m. Consequently, thankful people were able to work into the night.

Fortunately, the risk of fire was minimal. The cyclone was followed by a deluge of rain. The fire brigade extinguished a small fire that started before it could spread. Rain and mud complicated rescue work, but did not put a stop to it. ▪

Photo

The tornado was particularly devastating in one of the city's most elegant communities, where homes cost as much as $18,000. The cyclone also toppled huge structures such as the Winnipeg grain elevator. Smaller buildings fell like matchsticks in its wake. Tornado-damage estimates range from one to four million dollars.

Burns Plant Burnt in Blaze

Freezing Temperatures Hamper Firemen

JANUARY 13, 1913

Calgary, AB — In the city's worst fire to date, firemen faced nearly impossible conditions. Still, they made valiant attempts to contain the flames at Pat Burns' meat-packing plant. Nevertheless, the conflagration raged out of control for more than 36 hours.

Firemen worked in freezing temperatures which plummeted to -30 °F [-35 °C]. They were also hampered by an inadequate water supply. Despite a general alarm which rallied together all Calgary firefighters, the men had only six operational hoses. At times, only two of the city-owned hoses were operational because of problems with the water pump. Eventually, the city pump became virtually useless. At least a dozen hoses would have been necessary to contain the fire.

For the past six months, a controversy has been brewing over whether Calgary's water supply is adequate. The acting fire chief stated that water supplies were not a major factor in fighting the Burns plant fire. However, observers noted that water from the city hoses reached no more than 12-14 feet [3.6-4.2 m] above firemen while the flames rose two storeys high.

The mechanic in charge of the city pump claimed that the pump worked for 30 hours. A quick inspection has revealed debris, including deadwood and fish, caught in the screens of the pump. Fortunately, the packing plant's own pump and water cistern assisted in fighting the conflagration. Packing plant employees also aided the fire department.

Of undetermined cause, the fire began in the basement of the warehouse. Discovered by the night watchman, it eventually engulfed three packing plant buildings, fuelled by barrels of lard. The cold storage facility was one of the buildings destroyed.

As a result, losses include about 4,000 beef carcasses. Also destroyed were thousands of other livestock carcasses including hogs, poultry and lamb. Butter and egg stocks were also destroyed. Damage is estimated at about $2 million; however, the plant was fully insured. Burns Meat Packing is Canada's largest retail meat operation, and some are worried about rising meat prices. Already,

management is assuring the public that only minimal increases are expected.

In contrast to the financial loss incurred by this fire, Calgary fire losses were slight last year. The fire department responded to 380 fires, but damages totalled only $170,000. Such low financial losses are considered to be a record for a city of Calgary's size.

In 1909, the Calgary Fire Department was the first in Canada to use a motor-driven vehicle in department work. In the same year, the city hired 40 full-time firemen. Prior to that, volunteer firemen were paid 75 cents per call ▪

Calgary Fire Alarms
1905 - 1912

1905 – 44
1906 – 54
1907 – 93
1908 – 116
1909 – 125
1910 – 164
1911 – 274
1912 – 380

Photo

Midwinter fires often involve terrible conditions, as was the case with the Burns Packing Plant fire. Once clothing becomes wet in such icy weather, firemen become more susceptible to frostbite and hypothermia. Even after the flames have been extinguished, danger lurks. Then, firemen and clean-up crews enter unstable buildings. Inside, there is the risk that timbers encrusted with layers of heavy ice will give way.

Tragedy and Heroism

Disaster Claims 189 Lives

JULY 7, 1914

Hillcrest, BC — More than three weeks after the July 19 explosion at the Hillcrest Collieries, searchers have recovered one more body. One miner is still missing, but because of the severity of the explosion and damages, he may never be found. Including the missing man, the disaster has claimed 189 lives. Forty-six men were rescued.

To date the worst mining disaster in Canada and the third worst in the world, the Hillcrest explosion led to acts of unparalleled courage and heroism. Seconds meant the difference between life and death, and miners from other shifts worked relentlessly. When they didn't have tools, they used their hands to move rock, timbers and twisted mine equipment blocking the entrances. They mustered superhuman strength to free the hoist engine so that rescue cars could be sent underground.

One noteworthy hero was engineer Hutchinson. At the time of the 9:30 a.m. explosion, he was working near an entrance and was one of the first to reach safety. Before oxygen tanks and masks arrived in the government-equipment railcar, he and other Hillcrest miners went underground to look for survivors. Even when equipment and 100 miners from Blairmore reached the site, the Hillcrest rescuers didn't flag.

Quick-thinking mine manager John Brown saved many lives by sending immediately for an electrician. An unlikely hero to assist in the fallout of an underground explosion, the electrician restarted fans and reversed the wiring so that carbon dioxide, called afterdamp or black damp, could be drawn out of the mine. A mere 13 per cent of it in the air is deadly, and after the explosion, densities likely reached 50 per cent.

Two fires were discovered in the mine. In both instances, search-and-rescue teams came above ground, and men specialized in fighting mine fires went underground. Adding to the danger was methane gas, a highly combustible coal dust. Still, the firemen did their job, extinguished the fires and remained alert.

Finally, when no more survivors could be found, the miners recovered the bodies of their friends. According to the miners' creed, those who lost their lives must be brought back to the light, to their families and to a proper burial.

During the tragedy, three RCMP needed tremendous courage and dedication to perform their duties. Many police officers arrived to help, but Corporal Fredrick Mead and Corporal Arthur Grant had the most distressing job. They were assigned to the washhouse where the dead were brought. Many were dismembered, and for days, the officers reassembled victims, cleaned them and wrapped them in shrouds.

Constable William Hancock, assigned to collect evidence, accompanied search teams to the still danger-filled mine. He also helped recover the scattered remains of victims. Hancock's regular wage is 60 cents per day.

In the Crowsnest Pass, miners have resented police action, believing that it unfairly represents the interests of owners and a non-supportive government. Yet Hancock, Mead and Grant have more than proven themselves, and their special duties have garnered the officers the respect of miners and their families. On seemingly incessant duty, they worked in what seemed like a war zone, where death does not rest peacefully on the bodies and faces of its victims.

Women have worked tirelessly at the task historically assigned to them during tragedies. They provided food and coffee and helped look after the 130 widows and fatherless children, 400 of whom are under ten.

Some of the tragedies were particularly devastating. Miner David Murray stumbled out of the mine after the explosion and then realized that his three sons were still underground. Constable Hancock tried to prevent the disoriented man from returning underground until he had an oxygen mask and others could help him. Unfortunately, Murray rushed back into the mine, and neither he nor his boys came out alive.

Eleven years earlier, Charles Elick was in the mine at Frank when Turtle Mountain crumbled. Although he was buried, he was among the miners who dug themselves out. The Hillcrest explosion took his life. A day later, his child was born.

In the small community of Hillcrest, everyone has lost friends or relatives, and grief is pervasive. Just bearing it requires courage. ■

Photo

With the explosion, the roof of a hoist house collapsed. The mine entrance had to be cleared and the hoist repaired before a rescue car could be sent down the mine.

2,400 Homeless In Flood

Businessmen Show Heart

JUNE 29, 1915

Edmonton, AB — Fifty homes have been carried away by the raging North Saskatchewan River. An additional 700 houses are affected by the worst flood reported in Edmonton's history. About 2,400 people are homeless. Five businesses came to their rescue. These men own or rent about 275 vacant houses, and they have donated them for victims' use following the *Edmonton Journal*'s suggestion that vacant houses become part of the flood-relief program.

Others have also come to the aid of victims. Using rowboats to patrol, police rescued a man who was swept away by the current when he attempted to save his furniture. Two boys were also taken from the danger zone, and one man was rescued from the roof of his threatened home. Unfortunately, the river has claimed a young life. A baby died in the flood when its mother, trying to escape the swirling waters, dropped the child. Others were able to escape before their lives were at serious risk.

Those made homeless by the flood all lived in the river valley which divides the city. The community, however, is not divided, and almost everyone is contributing to disaster-relief in some way.

City Council's first official effort involved hiring all available trucks and wagons capable of moving belongings. Since then, the Board of Public Welfare has also established four relief depots. To help meet the immediate needs of the victims, those not directly affected by the flood have been asked to donate food, bedding and clothing at the downtown depot. The city has offered free water to those houses used as shelters.

Schools, the armouries and other public buildings have become shelters for people and their belongings during the crisis. Hotels and most houses donated as temporary shelter are vacant because of a decrease in population. In 1914, the city's population soared to 72,516, and many homes were built on speculation. Since then, Edmonton has witnessed a significant population decrease. The resulting availability of homes for victims is fortuitous.

In addition to homelessness, the flood has created other problems. Also at risk was the city's power plant, pumping station and water supply. At the power house, men have worked in water up to their waists

trying to maintain and restore service.

When the raging waters rushed around two local sawmills, lumber piles and log booms were swept loose. As a result, lumber has floated into the flooded areas and down river. The debris became an added danger for residents and rescuers. The Mill Creek Bridge was destroyed by floodwater and debris. Sixty of the homes swept away by the river were in the same area.

Currently, the North Saskatchewan River is moving at 16 miles [25.7 km] per hour. It crested at 2:00 this morning. Its high-water mark was 42 feet [12.8 m] above the low-water mark. At their worst, floodwaters were rising at about six inches [15.2 cm] per hour.

The last severe flood of the river was in 1899, when the water was 35 feet [10.6 m] above the low-water mark. Nearby communities along the North Saskatchewan River have also suffered extensive damage. ▪

Photo

Many buildings have been inundated with 5-10 feet [1.5-3 m] of water. Men returned in canoes and rowboats to the flooded areas to ensure that everyone had escaped and to save what belongings they could. Although warned of the impending state of emergency, residents of the river flats were slow to respond. Some left their homes with only the clothing they wore. Others piled what they could salvage on the high ground of the north side. There, piles were guarded by the women and children who had no place to go.

Heroic Horses Put Out to Pasture

Mechanization Claims Jobs

JULY 13, 1917

Vancouver, BC — They hurried to rescue many victims of raging infernos, but it is time to retire. The horses have had their day, and now they are being sent to pasture. Although Vancouver firemen are celebrating the complete mechanization of the fire department, they will miss their old partners in rescue work.

The horses have lived up to their duties. Over the past few years, they have been available as part of the team for significant fires. One fire broke out at 4:30 a.m. on April 29, 1915 on the centre span of the Connaught Bridge. For the firefighters, the day was bleak. One died fighting the blaze, and the collapse of the centre span plunged another into water 70 feet [21.3 m] deep below the bridge. Also that year, firemen had to attend three small fires on the Granville Bridge and one at the Hanbury Mill, but the heros of the past are no longer needed.

The horse-drawn steamers are kept at the stations just in case they are needed. The department is well equipped to deal with emergencies, but if the city were threatened, the old steamers would be pulled by motor trucks to the scene of a conflagration.

Forward-thinking Chief Carlisle had the city buy its first motor-driven fire equipment in 1908. At the time, he was subject to criticism, but his modern approach to fire fighting garnered honour for the Vancouver fire brigade. In 1889, he was appointed chief of the city's first paid fire department. Under his command were 15 full-time firefighters. They had an aerial ladder and depended on their horses to pull their heavy, modern equipment. That same year, the Vancouver firemen became North American champions in the hosereel race. To win, they competed against 12 American teams in Tacoma, Washington.

Soon, the city had water-hydrant service and 15 street fire alarms. Shortly after the turn of the century, the Vancouver Fire Department was chosen among the three top fire brigades in the world. The other two cities chosen were London, England and Leipzig, Germany. At that time, two downtown fire halls and nine local halls

served the city.

By 1911, the department consisted of 145 men, including one motor vehicle expert, 29 drivers and 65 who were horsemen or ladder men. As well as steam engineers, chemical engineers and linemen, Carlisle employed two telephone operators on the department's human team. ▪

Rescue Fails

Princess Sophia's Passengers Lost

OCTOBER 25, 1918

Vancouver, BC — For two and a half days in stormy seas, 268 passengers and 75 crewmen waited to leave their stranded ship. In the meantime, crewmen from rescue boats kept watch over the Canadian Pacific liner, the *Princess Sophia.*

On October 23, it left Skagway, Alaska for Vancouver. The six-year-old vessel was filled with passengers from Alaska and the Yukon, many looking forward to spending the winter in the south. Off course four hours after its departure, the ship struck the Vanderbilt Reef in the Lynn Canal.

The liner was stranded far enough onto the reef to appear dry and out of immediate danger. The winter storm was severe, and in the rough seas, the anchors of the rescue boats that arrived would not hold. Victims and rescuers waited for better weather so that the transfer could be accomplished safely.

As the weather worsened, the rescue boats took refuge behind nearby islands. The delays ended in tragedy. The last radio plea for help came when the ship began to take water. By then, it was too late. Swirling snow, poor visibility and rough seas meant that the rescue vessel that received the message, an American lighthouse tender called the *Cedar*, could not find the larger ship. No one on the *Sophia* survived.

Tragically, the lifeless bodies of hundreds of victims are being washed ashore and recovered by those who planned to save them. ▪

Photo

A 1916 fire had these fire department horses on the run. They pulled the only horse-drawn steam pumper still used in the downtown area of Vancouver.

Parade Turns Violent

Police and Protestors Injured in Fight

JUNE 21, 1919

Winnipeg, MB — Law and order have tentatively been re-established, but today became Bloody Saturday in the annals of Western Canadian history. Few can be labeled winners or heros after today's violent confrontation between RCMP and disillusioned citizens. What was to be a peaceful protest parade turned violent. One protestor is dead and more than 30 people, including 16 RCMP, were injured.

The parade was organized by workers and returned soldiers to protest the arrests of 10 strikers and the refusal of the federal government to act on unemployment and inflation. Those issues have been at the heart of the general strike, which began on May 15.

Perceiving the strike and parade as early steps in a Bolshevik-style revolution planned by the One Big Union, Winnipeg's mayor read the Riot Act. He outlawed the parade and called on RCMP to enforce his proclamation and to keep order. In sympathy with the strikers, most city police agreed to guard property but would not otherwise intervene.

Reports concerning the evolution of the violence are as yet conflicting. Protestors claim that without provocation, police opened fire on peaceful demonstrators. RCMP officials indicate that officers did not open fire until well after the protest had turned ugly. Strikers were armed, and when they shot at the police, officers' lives were in danger.

For RCMP, the day's duties began when 50 mounted policemen and 36 in vehicles were assigned to patrol Main Street, where strikers and sympathizers had gathered. According to police, strikers on rooftops threw bricks, cans, bottles and chunks of cement at them. The mob then rocked a streetcar, turned it over, set it on fire, tore off a side panel and hurled it at mounted officers. Strikers also pulled three Mounties from their horses and jabbed the horses with pocket knives and broken glass. Other officers did their best to help, but strikers attacked, clubbing both men and horses.

Ordered to do so, the Mounties fired a round into the air as a warning. When peace could not be established, they charged the crowd and fired shots at a lower angle to scare protestors. During the fray, Inspector Proby was attacked by two men, one aiming his gun at the inspector's head. In response, a corporal fired, killing him.

Strikers charge that police actions were unnecessary and escalated the violence. Most Winnipeg citizens have been in sympathy with the general strike, as up to 30,000 essential workers were involved. These included light, water, power, telegram, telephone and newspaper workers; garbage, postal, streetcar and some railway workers; bread and milk delivery men; and firemen.

However, when the shutdown began to effect every aspect of daily life, thousands of other citizens volunteered as relief workers. Hostilities intensified, resulting in today's confrontation. This evening the city is quiet as militia patrol the streets in vehicles with mounted machine guns. ▪

Photo

Horses were used in Riot Duty for the first time in Canada when violence erupted on the streets of Winnipeg. Almost all horses and mounted officers were injured. Dozens of strikers were also injured, and one is dead.

Mounties Lose Partners

Horses' Lives Claimed in Blazing Fire

OCTOBER 30, 1920

Brandon, MB — Today, in an early morning stable fire, many little-known heros who often performed above and beyond the call of duty lost their lives. Making every possible effort to save them, two Royal Canadian Mounted Police officers were injured. 31 of the Mounties' horses could not be evacuated before a blazing inferno claimed them.

Yesterday, flames raged through the old Winter Fair Building, which was used as an RCMP horse barn. When the doomed horses became frightened, they reared. Those trying to save them could do nothing.

One well-trained horse managed to save his own rescuer. Constable Larkin went into the stable for his mount. Wrapping the reins around his wrist, Larkin pulled the animal toward the door. The two were struck by a fireball falling from the roof. Then, despite the horse's serious burns, he dragged the unconscious Mountie from the fire. Sadly, the horse's burns have claimed his life.

The Brandon disaster is the second to affect the future of horses in the RCMP. This year, the riding school in Regina also went up in flames. The devastating fires have occurred when the RCMP is moving toward motorized patrols. Although they still use horses for some patrols, the Mounties now own 33 vehicles and 28 motorcycles. The history of man and horse in the RCMP is so important, however, that it seems likely that the two will always be linked in the memories of officers and in the public imagination.

The first Mounties, named the North West Mounted Police, came west in 1874. In that heroic march, the men's survival was dependent on horses, which suffered even more from exhaustion and starvation than had the men. In 1898, horses and men were again in a life-and-death situation. On Inspector J. D. Moodie's exploratory expedition from Edmonton through the Peace Country to the Yukon, all 30 horses died of exhaustion. Consequently, the dogs were fed the horses' remains, and the men survived.

From 1905 to 1907, the Mounties were assigned to build the Peace-Yukon trail. The all-Canadian road would link Northern communities. Again, Mounties suffered almost unbearable conditions. By September of 1906, many horses had died. That winter, some of the horses and men trekked south to

Peace River Landing in Alberta for the winter. The 19 horses and men had water only because the officers melted ice for both themselves and the horses to drink.

In the summer of 1907, the men built an additional 129 miles [208 km] of the 357-mile [575 km] Peace-Yukon road. In the course of this Herculean endeavour, another 26 horses died.

Even in less demanding conditions, the relationship between Mountie and horse has been crucial. Together they have patrolled vast, isolated areas in difficult terrain and bitter weather. Without the help of other partners, officers have saved their horses and mounts have rescued their Mounties.

The loss of the horses in today's fire at Brandon was deeply felt by the men who worked daily with the animals, and the legacy of their work for the Mounties and the country will not be forgotten. ▪

Photo

Just as Mounties are trained to lay down their lives if violence threatens the community, the horses of the RCMP have been trained to do the same if officers are in danger.

Climber Plummets to Death

Horror-stricken Wife Stranded 7 Days

AUGUST 5, 1921

Banff, AB — After more than two weeks, a search party has recovered the body of fifty-nine-year-old Dr. Winthrop Stone, following the tragedy that took his life. His wife, who was climbing with him at the time of the accident, survived seven days in the mountains before she was found by the search team. The party that rescued Margaret Stone was lead by a friend of the Stones, the renowned mountain guide Rudolph Aemmer. It included Bill Peyto, W. Child and RCMP officer Pounden.

On July 16, the Stones set out to make a first ascent of Mount Eon in the Mount Assiniboine group of mountains near Banff. When days passed and the Stones had not returned to their tent at the foot of Mount Eon, search efforts began. The search party included renowned climbers and CPR Swiss guides Rudolph Aemmer, Edward Feuz and Conrad Cain. All had become good friends of the Stones, having made many climbs in the Rockies with them. Although the rescuers were uncertain of the route the two had taken, the party set out up the mountain, using binoculars to scour the slopes as they progressed.

Just before dark, they heard a cry for help. Once he reached Mrs. Stone, Aemmer was able to climb down to the ledge on Mount Eon where she was stranded. It had been seven days since her husband fell to his death. When Dr. Stone reached a chimney 40 feet [12 m] above his wife on the mountain and another 60 feet [18 m] from the summit, he had unroped to explore. He either slipped or the rock gave way, and he plummeted 850 feet [259 m] to his death.

Horror-stricken, Mrs. Stone could not descend to search for his body until morning. Unsuccessful that day, she attempted to go down the mountain for help the next day but got lost. She dropped from her rope to

a ledge she believed might offer a route down the mountain and became stranded there. Without food or shelter, she scooped two holes to collect water to drink from the seepage of a crack in the mountain wall.

Nine days elapsed between the time of the accident and the rescue of Mrs. Stone. Aemmer carried the weakened woman on his back in a makeshift rope sling down the rugged rock walls to the timberline. Not until August 5 did searchers find the body of Dr. Stone, who was President of Purdue University.

In 1896, following the tragic climbing accident which took the life of Phillip Abbot, the CPR hired Swiss guides to assist tourists and mountaineers in the Rockies. Arriving in 1899, the first two of these guides were Edouard Feuz Sr. and Christian Haesler. Edward Feuz Jr. began guiding in 1903. Ernest Feuz and Rudolph Aemmer started in 1909. All have been involved in mountain search and rescue. According to Aemmer, "When somebody gets in trouble in the mountains, we go after him, take the necessary risks and bring him down. Nothing else counts." ■

Photo

This 1911 photo was taken when Edward Feuz (left), Rudolph Aemmer (right) and Dr. Stone (centre) were attempting the summit of Mount Lefroy. With the bad weather, they had to wait six days before they successfully reached the summit.

Two Pay Ultimate Penalty

Daughter Witnessed Policeman's Murder

MAY 2, 1923

Fort Saskatchewan, AB — Today, Florence Lassandro and rum runner Emilio Picariello forfeited their lives for the murder of Constable Steve Lawson. When he was shot, the Alberta Provincial Police constable was off duty and unarmed. Lassandro, the wife of Picariello's mechanic, fired the murder weapon. As an accomplice to the crime, Picariello was also sentenced to hang. Nicknamed "Emperor Pic," the rum runner was considered by the thirsty in Southern Alberta and B.C. to be a likeable, generous man and a respectable rum runner.

During the tragic murder and the gripping trial, it was the police officer's nine-year-old daughter, Pearl, who revealed the details of the crime. She witnessed her father's murder, and her testimony was essential for a conviction. The trial opened in Calgary on November 27, 1922.

Pearl had to stand on her toes to see over the witness box and point at Lassandro. Her testimony was straightforward. Old enough to understand the consequences of her words and to be aware of those who watched her, she said: "That's the lady who shot my daddy." After bravely giving her testimony, the little girl began to cry and was taken from the courtroom.

When her father died, Pearl was outside in the yard, waiting to be taken to the movies. A large car pulled up and the man asked to see her father. Pearl brought him. Then she watched as their talk evolved into an altercation with a gun. Shots were fired. Her father collapsed and the car sped away.

During the initial investigation, Pearl explained, "I saw him hold the man's arms over his head. The lady in the car fired shots, and Daddy fell."

When asked, "What lady?," she replied, "I dunno. Only she was wearing a scarlet tam."

As the APP, RCMP and British Columbia Provincial Police searched for the killer, those clues were key. During the trial, the testimony of 47 witnesses told the story.

Immediately, investigators suspected Picariello. Six large McLaughlin cars were used in the Emperor's Crowsnest Pass operation. Just prior to Lawson's death, he had been one of many officers in a car chase with Picariello and his nineteen-year-old son, Steve. At Coleman, with Steve speeding toward him, Lawson stood in front of

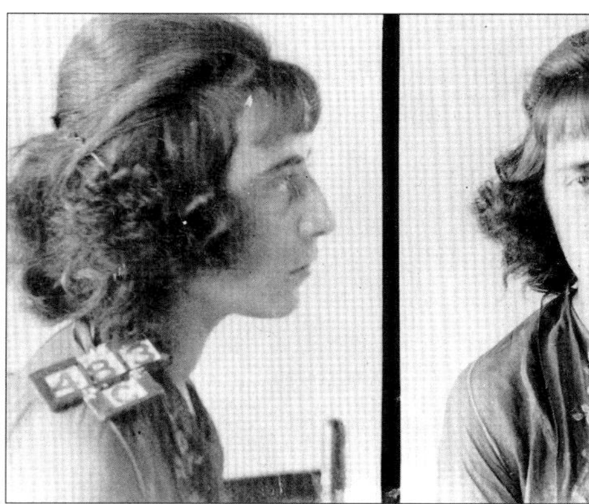

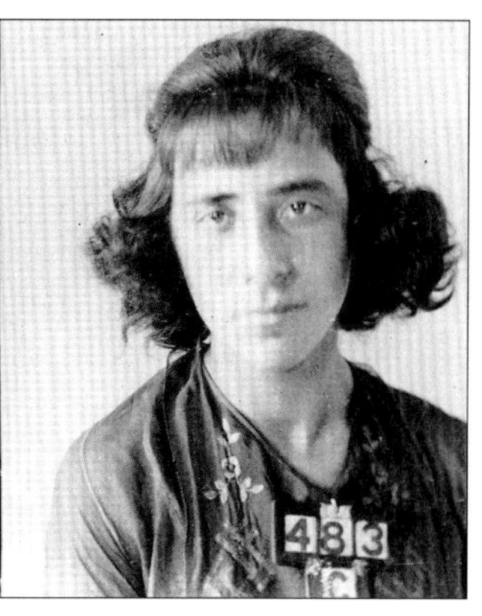

the car. Pic's son refused to stop, and Lawson fired two shots at the fleeing vehicle. Commandeering a car, Lawson gave chase and fired another shot, which grazed Steve's hand. Then Lawson's car tire blew and the fugitive escaped.

Unable to catch Steve, police turned their attention to Picariello Sr. During the chase, he attempted to prevent police cars from stopping his son's car, but Pic's vehicle carried no alcohol and he was not arrested. Inevitably, Picariello learned that Lawson had shot at his son.

Although he was generally a good-humoured gangster who did not antagonize police, Picarello threatened Lawson's life should the officer ever kill his son. Once off duty, Lawson returned to the Coleman barracks where he and his family lived. By nightfall, he was dead.

Investigators had no proof of Pic's involvement until a Blairmore taxi driver returned to town with Emperor Pic's McLaughlin. Police had the escape car with its bullet-shattered windshield. On the floor was a green, cloth-covered button.

The driver claimed that he had found the vehicle near a shack in the woods. The police search turned up discharged shells and a high-heel print. Eventually, the police flushed Pic from the woods and arrested him.

Picariello had heard that his son was wounded and was being held prisoner. The Crowsnest Pass had many small communities. Not knowing where the police had taken Steve, Pic decided that Lawson should help find his son. Twenty-two-year-old Lassandro and Pic drove to Lawson's home. The two men began to struggle with a gun. During that struggle, Lassandro fired and killed the constable.

Lassandro's green coat and red tam, the gun, shells and casings became evidence. Both were found guilty and sentenced to hang. Constable Lawson's daughter, Pearl, was courageous enough to tell her story to the police and the courts, and the justice system did the rest. ▪

Photo Left

At the time of his death, Constable Lawson was 35 years old and the father of five children.

Photos Right

Florence Lassandro mocked police when they went to question her. Once she was found guilty, thinking Picariello might be pardoned and she might be granted leniency, she signed a confession. She claimed to have fired in a state of panic.

300 Battle Mountain Blaze

Fire Claims Lives of Five Tourists

JULY 14, 1926

Kootenay Crossing, BC — The largest wildfire in Kootenay National Park's history has claimed four lives. Burning since July 6, the fire swept through the Vermilion and Kootenay River valleys in southern British Columbia near the Alberta border.

Because his family and friends were threatened, thirty-year-old CPR station agent L. R. Watt fought his way though the burning forest to get help. He reached the Kootenay Crossing depot and was able to describe the location of his friends and loved ones to rangers. While firefighters made their way to the threatened party only one quarter mile [402 m] away, the badly burned Watt was given first aid and taken to the hospital in Windermere. Unfortunately, help arrived too late for Watt's wife, two children, friend Clifford Nesbett and Nesbett's wife.

The Saskatchewan tourists were on a holiday camping trip. They were warned at the west entrance of the park not to go into the vicinity of the fire. Not realizing how quickly the flames could spread, they made their way along the Banff-Windermere Road into the threatened area.

Their tragedy unfolded when they rounded a bend where the fire was burning. The car stalled. Believing they could return to safety, the men turned it around by hand. By then, the group was surrounded by fire and Watt set out on foot for help.

When firefighters reached the scene, they found the twenty-seven-year-old school teacher unconscious. He was sprawled over the body of his twenty-five-year-old wife, as if to extinguish flames or to protect her. Shortly afterwards, he died in the Lake Windermere Hospital. The fire also claimed the lives of Mrs. Nesbett, Mrs. Watt, the Watts' twelve-year-old son and their ten-year-old daughter.

Undoubtedly hoping to escape the fire and heat, the victims had tried to dig their way into the road bank. When the gas tank of the car exploded, their fate was sealed. The explosion sprayed them with lighted gas.

A lightning storm in the Vermilion and Kootenay River area on July 6th is suspected to have caused the fire. The timber which stood for 15 miles [24 km] along the Windermere road was destroyed by the fire. Smaller burns have also caused significant damage

to other areas. To fight the fire, men were recruited from as far away as Calgary.

Kootenay National Park, the site of the fire, was established in 1920. Three years later, the Banff-Windermere Road was completed and work began on a chain of campgrounds in the park. During the 1924 tourist season, 4,500 automobiles traveled the highway where the Watts and the Nesbetts were to lose their lives. ■

Forest Fire Causes in Canada, 1925	
Campfires	*944*
Lightning	*978*
Unknown	*714*
Railways	*711*
Settlers	*692*
Smokers	*531*
Misc. known	*431*
Industrial	*257*
Incendiary	*204*
Public works	*28*
Total	*5,490*

Photo

Losses during this year's firestorm in southern B.C. accentuate the need for forest fire prevention and effective fire fighting equipment and techniques. Canada's first Save-the-Forest Week was established in April, 1923. In 1920, this concept was initiated in the USA. For the 1926 campaign, Canada and the USA agreed to the same week as a focus for their activities.

Explosion Rocks Town

Latest Safety Equipment Used

SEPTEMBER 22, 1926

Hillcrest, AB — Three days after yet another coal-dust explosion rocked the Hillcrest Mine, rescue workers found the second of two men who lost their lives. No others were in the mine at the time. Within minutes of the explosion, the rescue party arrived with the latest in safety equipment. Like most rescue parties today, some had first aid training. Soon, they found the first miner. Today, they found the second. This time, no amount of equipment or training could have saved the unfortunate miners.

Given the tragic history of the Crowsnest Pass, rescue teams there are extremely well prepared. Because of the daily risks, miners have embraced the concept of emergency first aid training. Colliery doctors are available, but can only help if the victim is breathing.

Many B.C. workers whose lives can be in danger due to a simple mishap are also enrolling in emergency aid courses. Current trends and the official statistics for 1920 indicate that B.C. and Nova Scotia are the only provinces in which enrollment in St. John's Ambulance Association first aid courses is still on the rise.

In 1911, all four provinces in Western Canada established branches of the St. John's Ambulance. Already in 1915, the first aid and rescue competition of the Crow's Nest Pass Coal Company attracted 11 teams, totalling 55 men and 10 boys. That year, 139 of the company's employees took the St. John's examination. As a result, the mine had one person trained in first aid for every 14 men underground. Today, in much of Western Canada, the association has not only ambulance and instruction divisions but fire brigade and nursing divisions.

Another organization that has brought reassurance to isolated, logging and mining areas is the Victorian Order of Nurses. In 1898, the VON began opening hospitals in remote areas. The society has founded hospitals in such B.C. communities as Revelstoke, Kaslo, Fernie, Barkerville and Windermere. Like other rescue workers, the VON are prepared to travel

wherever they are needed and by any means available: on foot, on horseback, by automobile, train, wagon, cutter, bicycle or motor scooter.

In 1920, a sister association—the newly organized Canadian Red Cross Public Health Nurses—began visiting lighthouses and rural districts in B.C. Prevention of illness, injury and disaster will always be the best option, but at least communities and individuals can rest assured that in times of need, help will be on the way. ■

Photo

The 1920s have delivered the first motorized ambulance to Queen Victoria Hospital at Revelstoke. The VON raised money for the hospital, which opened in 1902. It replaced an eight-bed facility that operated from a house.

S.S. *Prince Rupert* Strikes Rock

Series of Disasters Plagues Ship

AUGUST 22, 1927

Vancouver, BC — Today's accident of the S.S. *Prince Rupert* was one more in a series of disasters that have plagued the ship. This time, the ship struck Ripple Rock in Seymour Narrows. Fog and strong tides contributed to the accident, and the nearby cliffs further complicated rescue efforts.

One of the ship's propellers was raised out of the water and the starboard propeller collapsed: the *Rupert* was in a very dangerous position. Bad weather made matters worse. Luckily, the S.S. *Cardena*, which carries mail and cargo, was nearby. Despite risks to himself and his crew, Captain Andy Johnson manoeuvred the S.S. *Cardena* alongside the damaged ship. A towline was rigged between the two vessels. The *Cardena* pulled the *Rupert* off Ripple Rock and to the safety of Deep Cove. Then, the *Cardena* took aboard as many passengers as possible while the rest were rescued by the *Princess Patricia*, a CPR ship. The passengers of today's accident are safe.

While plying the B.C. coast from Vancouver to Alaska, the *Rupert* has had many accidents. The most recent one occurred seven years ago on September 20, 1920 at Swanson Bay. Before that, in January of 1919, she was sitting out a thunderstorm at a port in Vancouver. Suddenly, her mast was struck by lightning. No one was hurt, but the lightning split the mast and immediate repairs were necessary.

On March 23, 1917, the *Prince Rupert* was hit by gale winds of 70 miles [112.6 km] per hour. In poor visibility, she left her northern port of Prince Rupert. She was traveling to the small inlet community of Anyox. The ship was still close to port when it ran onto the rocks of Glenn Island. As the accident happened only 30 feet [9 m] from the heavy forest, passengers were able to walk ashore to safety. No one was hurt. The *Rupert* signalled for help and rescuers picked up passengers.

This time, the ship was in serious trouble. Rock had to be blasted away before rescuers could pull her out to sea. Eventually, they towed her to port for repair, and soon the *Rupert* was back on the seas. With such a dramatic history, the *Rupert* may make prospective passengers wary. ▪

Photo

Seven years before the 1927 accident, the S.S. Prince Rupert *sank here in Swanson Bay. Fog had created hazardous conditions, and as the* Rupert *traveled toward the pulp mill port, she struck a reef off Princess Royal Island. Fortunately, the captain was able to beach the ship before the passengers were at serious risk, and there were no injuries. However, there was a 12-foot [3.6 m] tear in the bottom of the ship. As a result, when the tide came in, the water caused the* Rupert *to keel. By high tide, about 72 feet [30 m] of water almost entirely submerged her stern.*

Pilot Pulled From Wreck

Nose-dive Spells Disaster

APRIL 25, 1929

Calgary, AB — An airplane accident has claimed the life of H. E. Edwards, who died in hospital this morning. His plane crashed at 4:40 p.m. yesterday. The pilot had just left the airport, 1-1/2 miles [2.4 km] away. Near a ravine on the Old Banff Coach Road, the Gypsy Moth suddenly went into a nose dive.

From an altitude of 200 feet [61 m], it hurtled to the ground in a crash that could be heard as far away as the airport. The force drove the plane's propeller two feet [0.6 m] into the soil and pinned Edwards into the cockpit.

Despite the danger of a gas tank explosion and the difficulty of prying loose the mangled steel, passenger Steve Soltice and a farmer working close to the crash site did everything possible to save the pilot. Witnesses from Great Western Airways also rushed to the scene to help. However, when pulled free of the crumpled fuselage, Edwards was unconscious.

As soon as possible, he was taken by ambulance to the Calgary General Hospital. Edwards suffered from internal injuries, his limbs had been crushed and his face had deep gashes. Recovery was out of the question.

Soltice, the passenger in the plane, was fortunate. He was treated for cuts to his legs and then released from hospital. He is currently a student of the Calgary Aero Club. Edwards was also a member of the club but had not received instruction at the flying school. However, he was one of the first to pass the new government tests for a private pilot's license.

A week earlier, the Calgary airport was the site of another crash. This time, the more fortunate fliers walked away uninjured. The pilot had been about to land, his plane only 30 feet [9.1 m] above the airfield. The youngest pilot working for Great Western Airway had tried to stop too quickly, throttling the engine below flying speed. The plane dropped and the undercarriage collapsed, but the pilot and his student passenger were uninjured.

Edwards' tragic accident occurs at a time when Calgary has become increasingly committed to the development of aviation facilities. The city is planning to add four hangars to the new civic airport. Pilots have long known that winds in the Calgary area create risks, but most are adamant that airways will be the high-

ways of the future. They predict that planes will carry more than mail and special passengers. Many believe that aircraft will be used increasingly as air ambulances to fly residents of small, isolated communities to medical facilities.

Western Canadians are at the forefront in adopting aerial transportation. In 1928, the Canadian government surveyed privately owned aviation facilities, aircraft and personnel. As would be expected given their much larger populations, Ontario and Quebec have more of each. Considering the West's relative population, however, it has proportionately far more planes, pilots and airfields. ■

Photo

The Gypsy Moth was delivered to Calgary's flying club just a week before the accident. That Sunday, its test flight proved successful. Edwards was a promising pilot, and his experience far exceeded the 10 hours of solo flying required to carry passengers. Calgary winds can be perilous and were likely responsible for the accident.

Manitoba Fights Forest Fires with Airplanes

Old War Planes Used for Fire Duty

JULY 15, 1930

Winnipeg, MB — Will the West continue to save its forests from the skies? The question now faces prairie provinces, which are winning control of their natural resources. Today, Manitoba became the first in the West to gain full power, control and responsibility with regard to forests and fighting forest fires. By the end of the year, the other prairie provinces are expected to have the same rights.

Since 1919, brave Royal Canadian Air Force pilots have stepped into old war planes to perform fire patrol duty. That approach resulted from co-operation between the Canadian Forest Service and the RCAF. In 1928, 24 airplanes patrolled the prairies.

Although the view was great, the job had its disadvantages. When a pilot went down in any of the isolated, forested areas, finding him was difficult, even for competent search-and-rescue teams. But the RCAF pilots were assigned the job, and they flew military planes to do the work.

With the new Natural Resources Acts, the RCAF is being withdrawn from the patrols. The federal government is no longer responsible for forests, except in the national parks.

Instead, each province will decide whether to continue aerial patrols. If they choose to continue the patrols, they must assign departments to do the work or contract aerial patrollers.

Already, Manitoba has a strong record with respect to fire loss. In 1876, it was the first province to establish a provincial fire office and Fire Commissioner. With its Fire Prevention Act of 1912, Saskatchewan became the second western province to pass legislation regarding its responsibility for fire fighting and fire prevention. In 1912, B.C. organized its own Forest Service, with responsibilities related to forest fire control. Alberta has passed some forest-related legislation but has not established provincial officials or a department dedicated to forest and fire services.

Last year, forest fires plagued Western Canada. With little humidity and high winds, the prairies and B.C. suffered serious losses. Given drought

conditions, this trend is expected to continue in the 1930s, and aerial fire patrols seem to be a good option. Many expect the western provinces to absorb the programs and personnel of the Dominion Forest Service.

At present, planes are used primarily for detecting forest fires rather than suppressing them. In the forests, landing areas are unavailable. Even if a pilot were able to land, the small planes can carry limited equipment and few men.

By plane, a Chief Ranger can quickly survey a fire site. The sooner the site can be assessed and a fire-fighting strategy developed, the more likely it is that small fires will be extinguished before they rage out of control. ▪

Fires in Banff National Park Year / Hectares Burned
1880 – 89,637,050
1890 – 99,918,600
1900 – 91,216,050
1910 – 19,63,300
1920 – 29,910,950

Photo

This plane patrols northern Saskatchewan. The province has embraced the new aerial methods of fire detection to protect its forests. Planes such as the HS2L and the Gypsy Moth have already proven their worth in the fight against forest fires.

Winning Combination

Cops and Community Combine

JANUARY 27, 1931

Edmonton, AB — The statistics are out. Clearly, communities want policemen. However, the latest report by the Mayor of Edmonton has people wondering exactly how many police officers are enough. Edmonton has one officer for every 331 acres [134 hectares]. Winnipeg has one for every 57 acres [23 hectares], while Vancouver has one for every 83 acres [33.5 hectares]. From a slightly different perspective, Saskatoon has one officer for 1,024 inhabitants. Edmonton has a ratio of one to 917, Regina's is one to 916 and Vancouver's is one to 618.

Most Edmontonians would like these figures to prove that their department is among the best in Western Canada, but the report also indicates that the average age of its officers is 47 years and 146 days. According to the mayor, this is too old for the officers to have much "pep."

Whether or not they have pep, there is a history of progressive thinking in the Edmonton Police Department. The school patrol program is only one example. With that program, Edmonton police acknowledge that they must do more than track down criminals and prevent property damage. Working with the community is a key to policing success.

Historically, other progressive measures have included appointing the first full-time female constable and first full-time Native constable in Canada. As far back as 1909, Alex Decouteau, who grew up on the Red Pheasant Reserve in Saskatchewan, was made a constable. Shortly after moving to Edmonton, he joined the police. Many Metis and Native people had worked as police scouts or as special constables on the reserves. However, Decouteau assumed exactly the same duties and responsibilities as other officers. While working for the department, he was promoted to Sergeant.

While he was an Edmonton police officer, he demonstrated his pep. An outstanding middle-and long-distance runner, Decouteau was the only Albertan on Canada's Olympic team for the 1912 Games in Sweden. During the First World War, he joined the army. Sadly, in 1917, during the Battle of Passchendale, Edmonton lost its Native police officer and world-class athlete.

Annie Jackson's appointment was another first. She was hired in 1913 and was Canada's first full-time

female constable. Her primary duty was working with women and girls in conflict with the law. That same year, Los Angeles claimed the first female police officer in the world. However, in the years prior to that, the Edmonton police chief had requested that Mrs. Emma Robinson be hired as a full-time officer. When that request was denied, she remained as a part-time matron for the department.

Edmontonians are likely to support whatever actions will keep the department progressive and the community safe. ▪

Monthly Wages — 1931

Police Department Edmonton, Alberta

Accountant	*$185*
Sergeant - Detective	*$170*
Detective	*$165*
Sergeant	*$165*
Acting Detective	*$150*
Constable - 1st Class 7 years experience	*$150*
Constable - 1st Class	*$145*
Constable - 2nd Class	*$135*
Constable - 3rd Class	*$125*
Janitor	*$120*
Policewoman	*$100*
Matron	*$97*

Photo

Police officers have long been heros to children. In 1928, with Jasper Avenue and 105 Street becoming an increasingly busy intersection, Constable R. Robertson added school patrol to his other duties. More and more automobiles are coursing down city streets, and some drivers are not slowing their speed near schools and play areas. Children can no longer depend on a good horse to wait as they cross streets. School patrol is considered to be preventative policing and public relations work. The officer teaches children to be alert around traffic and demonstrates that policemen are like concerned parents who expect their kids to respect the law.

Regina Rocked by Riot

Police Pitted Against Strikers

JULY 2, 1935

Regina, SK — Tear gas, sticks, rocks, riding crops, bricks and bullets became ammunition in the three-hour riot that erupted on the streets of Regina last night. City detective Charles Millar was killed and RCMP Constable Wakefield is in serious condition with head wounds and a concussion. At least two strike leaders were seriously injured. One is in hospital with a serious side wound and the other was picked up by rioters and has disappeared, so his condition remains unknown.

During the riot, about 100 people were arrested, eighty of whom are being held in RCMP or city police cells while another 20 are under guard in hospital. The two city hospitals helped about 100 police and rioters who needed attention for injuries. However, because of the threat of arrest, some strikers are likely tending their wounds in private homes.

Today, Saskatchewan's Premier Gardiner put into effect a plan to feed the 1,200 to 2,000 people who became riot refuges at the Exhibition Grounds. Many of the men in the stadium have not eaten in 24 hours. Food will be provided until they can be peacefully dispersed to their homes, camps or places of origin.

Following the confrontation, protestors fled to strike headquarters at the stadium. Armed with rifles, RCMP were placed on guard at the building, and men have not been allowed to leave except as part of the delegation to see the Premier.

The Premier wants the federal government to withdraw the RCMP from strike-related duties in the city and return authority for policing and justice to the province. In mid-June, Prime Minister Bennett was informed that western strikers from as far as B.C. were staging an On-to-Ottawa trek. Their complaints included starvation conditions at the relief camps, and many of the leaders wanted a work-for-wages program.

Believing the agitators to be communists intent on overthrowing the government, the Prime Minister ordered that strike leaders be arrested. The cabinet passed an order-in-council requiring the RCMP to stop protestors at Regina. Since then, the 2,000 riding the rods have been detained at Regina camps. To prevent them from reboarding trains, about 500 RCMP and CPR police have been guarding the station and camps.

Yesterday's riot erupted after a mass meeting,

staged as a send-off for protestors continuing the trek. By about 8:15, city police attempted to dispel the crowd of men, women and children. When the mood between police and strikers turned ugly, the RCMP responded. After a fierce battle in which many people were trampled and injured, strikers fled to their homes or to the stadium.

In Winnipeg yesterday, protesters were determined to resume their trek. About 10,000 strikers and curious onlookers gathered for a meeting and send-off. Speakers raised fears of being stopped and thrown into concentration camps. By about 9:00 p.m., those continuing the trek started for the rail yards. Reacting on their fears of a police trap, they took possession of the city's soup kitchen. They demanded three meals a day until they were allowed to continue the trek without the threat of imprisonment. The protest did not become violent, but tensions were very high.

Although Regina's riot is the worst to erupt in Western Canada since 1919, workers throughout the four provinces have demanded food aid, shelter and employment. RCMP and city police, who formerly helped citizens in times of disaster, have been ordered to stop the demonstrations. Some officers and many citizens sympathize with the plight of the unemployed and their families.

However, civil and federal authorities have pitted the police against the poor. This spring on April 23, about 2,000 protestors fighting with police caused thousands of dollars in damage to the Hudson's Bay store in Vancouver. In 1932, 16,000 attended a hunger march in Edmonton. Mounted police rode into the crowd, but no one was killed. On September 29, 1931, three miners were killed by RCMP during a strike at Estevan, Saskatchewan. Without government action concerning employment, relief camps and farm aid, the crisis may have ended still more violently. ▪

Photo

Detective Millar (centre, on the ground) died after being hit in the head with a club. Strikers continued to advance as an officer tried to help Millar.

Gunmen Murder Four Mounties

Brutal Killings Shock Community

OCTOBER 9, 1935

Banff, AB — Yesterday, RCMP, local police departments and the national park services joined forces to end a murderous rampage. In it, four policemen lost their lives. Neish, a park warden and former RCMP officer, was one of the capable and heroic men who brought the tragic story to a close.

Two officers were murdered near Benito on the Saskatchewan/ Manitoba border. The other two policemen were shot about five miles [8 km] east of Alberta's Banff National Park. This criminal episode took the lives of more Canadian lawmen than any other single incident to date.

At the scene where his two colleagues were fatally injured was Constable Combe of the Banff RCMP detachment. He opened fire, killing Joseph Posnikoff, the group's ringleader. During a standoff with the other two killers, Neish called for their surrender. When they replied with a barge of bullets, crack-shot park warden Neish took down John Kalmakoff and Peter Woiken.

The three responsible for murdering the officers were Doukhobors, a Pacifist Russian religious sect. The young men's violent actions and drinking were totally incompatible with the Doukhobor commitment to pacifism and temperance.

The first to die were RCMP Constable John Shaw and municipal officer Constable William Wainwright, who suspected that the young men were responsible for a shop-breaking crime in Saskatchewan. On Friday, October 4, the officers ordered them into an unmarked police car. They drove toward Pelly, Saskatchewan, 20 miles [32 km] away. There, the suspects were to be interrogated, and if necessary, jailed.

On the way, the officers were overpowered. Using a knife to attack them and wrestling away a police revolver, the three assailants brutally killed the officers. The perpetrators stripped the policemen of identification and valuables, dumped them in a slough and escaped in the car.

By Monday night, the three were sighted in the Exshaw and Canmore areas. The park gateman at Banff had refused them entry into Banff National Park when they could not pay the $2.00 entry fee and would not answer questions about the their car. Shortly after reporting the suspicious men, he again phoned the Banff detach-

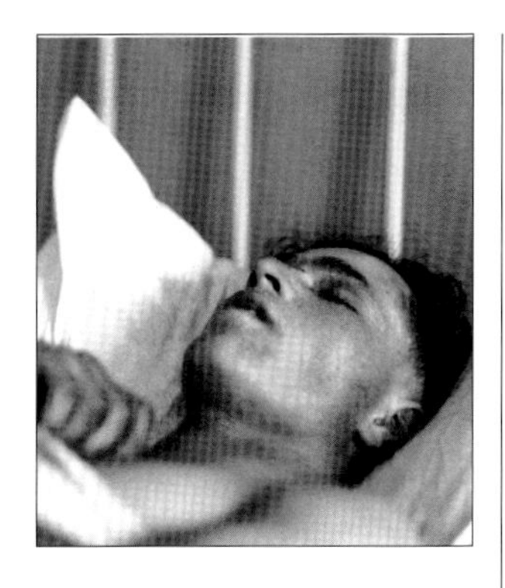

ment. A visitor reported that a couple had been robbed at the roadside east of the park.

With that news, officers from the Banff detachment drove east of the Banff park gates. Two off-duty officers, Sgt. Wallace and Constable Combe, accompanied uniformed officers Grey Campbell and George Harrison. Just east of the park, they angled their car to block the road.

The first vehicle to stop was a Calgary couple. En route to Banff, they noticed a stranded vehicle and stopped to offer help. They were robbed, and although they feared for their lives, they were released. The bandits, they claimed, were traveling behind them. The couple was waved through the roadblock.

Next, the fugitives' car stopped. In the glare of its headlights, Wallace and Harrison approached. Just before they reached the bumper, they faced a blast of gunfire. Wallace was hit in the throat, but before dying, he shot out the car's lights. With a bullet lodged in his chest, Harrison called for backup. The other officers returned fire.

When help arrived, Campbell took the injured men to the Canmore Hospital. Later, despite being transferred to Calgary for specialized medical attention, they died. Meanwhile, Combe, another constable and a magistrate continued to fire at the wanted men. Three shots from Combe struck and killed twenty-year-old Posnikoff. In the darkness, the other two escaped into the bush.

Although the weather turned ugly and morning brought a swirling snowstorm, the manhunt proved to be short. The area was cordoned off. Cars and trains were searched. Volunteers and police combed the area. The new snow revealed tracks. The police dog, Dale, picked up the scent of the fugitives.

When his group of searchers was fired upon from the bush, Warden Neish called for the men's surrender. Their reply was another barge of bullets. Then, the marksmanship of Neish ended the tragic story. ▪

Photo Left

Injured in the final shootout, eighteen-year-old Peter Woiken (above) was taken to the hospital in Banff, where he died. Twenty-one-year-old John Kalmakoff also died in hospital.

Photo Right

Game Warden Neish was stationed at Mosquito Creek, northwest of Lake Louise. When he heard the news on his park radio, he joined the search. He recorded his actions in his daily journal, which was sent to headquarters in Ottawa. For October 8, his entry was simply, "Killed two bandits."

Pier Inferno Worst Yet

Residents Demand Another Fireboat

JULY 27, 1938

Vancouver, BC — Pumping water from a raft beneath Pier D, firemen once again risked their lives on the Vancouver docks. The conflagration, which raged beyond control, broke out this afternoon on the Canadian Pacific Railroad's Pier D. The pier was built by the CPR in 1920, and damage is expected to be in the area of $1 million.

As the dock was situated at the foot of Granville Street, huge crowds of spectators gathered. Once the fire started, the crowds made vehicle access to the area almost impossible. Not only did they impede access, but they endangered themselves and complicated the work of firemen.

No lives were lost, and despite the crowds' being in the way, many of the spectators grew anxious about the special risks confronting Vancouver's firemen. Working beneath the docks seemed to be very dangerous.

The piers are old and grease-covered. Their wooden warehouses offer readily available fuel. Lumber and forestry products and the fuel needed for the shipping industry are frequently stored there, and all are highly flammable.

As a result, this latest fire has already provoked considerable protest. Once again, residents of the city are demanding another fireboat. In the past, the Harbour Board has simply ignored the plea.

In 1928, private industry commissioned the building of a fireboat, the *Carlisle*, for exclusive use on False Creek. The boat has provided excellent service, but with the expanding dockyards, increasing numbers of people near the docks and expanded storage capacity needed for shipping, port citizens feel that another fireboat is essential. ▪

Photo

The Pier D fire occurs during one of the city's hottest summers. As clouds of smoke began to billow around the dock, the alarm went off. At the time, smoke was already drifting to Vancouver from a devastating forest fire on Vancouver Island.

Gas Attack?

Soldiers Decontaminate Streets in Simulated Demonstration

SEPTEMBER 21, 1940

Edmonton, AB — Yesterday, city streets were filled with a wartime gas-decontamination squad. With its siren screaming, a fire department car careened though the streets and raised the alarm. From somewhere, a white, smoky cloud began to filter though the air.

Next came the anti-gas tank truck. From the truck, men hosed the curbs, flushing water along the street to clear Jasper Avenue and 105 Street of tear gas. Other soldiers swept gas from the drenched streets while Edmontonians wiped their eyes with damp handkerchiefs.

The gas attack was merely an exercise that hundreds came to watch, but the soldiers and their methods were the real thing. Most were from the Canadian Small Arms Training Centre in Calgary. Looking like men from Mars, they wore respirators, helmets and gas-proof capes during the feigned air raid. A smoke screen simulated tear gas, but some genuine tear gas was used in the demonstrations. Although it caused some tears, the negligible amount was harmless.

Later, Clark Stadium was the scene of other battle-related demonstrations. Units from Edmonton, Calgary, Vancouver and Seaforth illustrated modern methods of battle. The huge, appreciative crowd that attended was more curious than fearful. Few can imagine their city as the target of an attack.

Vancouver, however, is taking civil defence exercises far more seriously. As a major seaport, industrial centre and terminus of two national railways, that city could be the target of an air raid. Residents fear incendiary gas and high explosive bombardments by the Japanese. In fact, Vancouver's city council wants the federal government to add anti-aircraft defences to the coastal guns, lighthouses and navy patrols. Some also want torpedo nets to protect piers.

The port city is seriously considering issues of civil defence and ways of providing residents with adequate help in case of a city-wide emergency. An Air Raid Precautions Unit has been organized. It operates according to the English system of appointing wartime fire wardens who are trained to carry out their duties during a state of emergency. Vancouver is also planning a city-wide blackout in the spring as part of its defence plan. ■

U.S. Military Invades Edmonton with Shovels

Americans Have Proven to be Friends

NOVEMBER 16, 1942

Edmonton, AB — The American military machine, in the form of Alaska Highway construction workers and American Air Force personnel, has come to Edmonton's rescue. Yesterday, the city was crippled by a storm that dumped 19 inches [39 cm] of heavy snow onto the streets.

The USA chose Edmonton as the largest northern city from which they might direct the construction project. The American Air Force also needed a stop-over for refuelling and maintenance. As neighbors, Canada and the USA became committed to building a highway that would benefit both countries. As allies, they are mutually supportive in terms of defence plans. As a result, thousands of American men have been stationed here.

The Edmonton disaster had its beginnings late on Sunday, November 15. During a heavy snowfall, cars and streetcars became stranded. By Monday morning, the rest of the vehicles in the city were buried or blockaded by snow and drifts. The worst drifts were up to 15 feet [4.6 m] high. As a result, in many car lots and residential areas, vehicles were completely buried. In other locations, only the tops of vehicles were visible.

Conditions paralysed businesses and closed schools. Of even more concern was the fact that the sick, injured, needy and elderly might not receive the attention they need. However, like most other cities, Edmonton has no snow-removal equipment. Little could be done to alleviate the situation.

With their Edmonton hosts in trouble, the Americans decided to dig out the city, providing disaster relief in the form of labour. They also brought their scrapers, trucks and loaders to clear streets and recover buried vehicles.

Now that the job is nearly done, highway construction personnel will return to their task of planning and building a road to link Fairbanks, Alaska and Dawson Creek, B.C. It will cross more than 1,500 miles (2,575 km) of some of the world's most difficult terrain. The road will make it possible to transport supplies, equipment and men to Alaska in case of a Japanese invasion.

The air force will also return to military priorities, including the defence of Alaska and the lend-lease program that exists between America and Russia. In the meantime, on Edmonton's home front, the Americans have proven to be friends. ▪

Photo

One of hundreds of vehicles used in the construction of the Alaska Highway, this truck rolled down a bank north of Fort Nelson, B.C. The American highway builders have had extensive experience rescuing people and recovering vehicles when bad weather has created treacherous conditions. Not only has progress been hampered by snow and rain storms, but workers have faced mud slides and floods while working on the roadbed.

Ship Explodes

Bravery Averts Further Disaster

MARCH 9, 1945

Vancouver, BC — After burning for three days, the S.S. *Greenhill Park* fire has been extinguished. The disaster leaves eight men dead. Without the heroism of tugboat Captain Harry Jones, his crew and the crew of the *Greenhill*, the entire Vancouver harbour might have become a disaster area.

In the midst of the conflagration, the captain and his crew managed to tow the burning freighter away from a Vancouver wharf. The ship exploded at Pier B on a Canadian Pacific dock after a small fire broke out in one of the holds of the ship. Longshoremen and painters who had been working on the ship fought the fire.

Just before noon, the fire caused the explosion that claimed the lives of the eight men. It blew out hundreds of windows in the downtown area and flying debris injured dozens of people. Flames and smoke rose from Pier B at the foot of Burrard Street, signaling the disaster on the horizon.

The city fireboat tackled the flames, but the fire was beyond control. Worry mounted concerning the possibility of another explosion, which might be worse than the first.

The ship was a waiting time bomb. The cargo aboard ship included 50 barrels of whiskey, 94 tons of explosive sodium chlorate, 2 tons of calcium silicide and more than 7 tons of lifeboat flares.

Captain Jones and his tug, the RFM *Kyoquot*, had been towing a barge to the B.C. Sugar Refinery when he was alerted to the fire. Aware of the enormous threat it posed to life and property and the difficulties of fighting ship and dock fires, the Captain cut loose the barge and made for the fire.

Once at the site of the fire, the tug's captain and crew threw a towline to the ship. Many brave men remained aboard the burning *Greenhill* to make sure she could be towed away from the dock. Despite the incoming tide, the tug pulled the freighter three and a half miles [5.6 km] from the Vancouver wharf. Beached at Siwash Rock at the extreme end of Stanley Park, the ship no longer posed a threat to the city.

Although the *Carlisle* continued to pump water on the beached ship, it could not be saved. Only the hull of the ship remains. Officials believe that the unacceptable manner in which cargo had been stored contributed to the disaster. The combination of explosive cargo and careless storage are likely to have caused the explosions. It is believed that a longshoreman may have

dropped a match. Whiskey had leaked onto the deck, perhaps as a result of human carelessness. Once the alcohol caught fire, it carried the flames to the other cargo. As well as the other flammable and explosive materials, the government-chartered freighter was loaded with lumber, airplane parts, cotton and cloth. To date, the fire is considered the worst in Vancouver since the city burned in 1886. ■

Photo

The fireboat, the Carlisle, *and another boat pulled the giant S.S.* Greenhill Park, *but they could do little to extinguish the fire.*

Captain Rescued

Final Disaster in Ship's Eventful History

SEPTEMBER 22, 1945

Ketchikan, AK — Captain Neil MacLean saved passengers and all but one crew member when he beached the burning S.S. *Prince George*. The Canadian ship had set sail from Ketchikan when one of its fuel tanks exploded. A man in the engine room died in the explosion.

Despite the fire spreading through the ship, the captain remained levelheaded and in command. He beached the *Prince George* on a rock in the harbour. The US Coast Guard came to the rescue of passengers and transported them ashore. Next, the rescue vessel towed the burning ship to the nearby Gravina Island. There, the fire could burn out without further endangering life and property.

Only then was Captain Maclean ready to leave his ship. Given the danger in approaching the burning vessel, the Coast Guard expected the captain to jump overboard and swim to the rescue vessel. The captain could not comply.

Although he had been a seaman for about 35 years, Captain Maclean couldn't swim. To save him, rescuers had to manoeuvre close enough so that the mariner could step from the rail of his own ship to the safety of the Coast Guard vessel.

Because of the extent of this 1945 fire aboard the *Prince George*, the ship's career is likely over. While no other lives have been lost on the ship, luck has not always been with her. The northern waters and freezing weather have proven to be treacherous.

Eight years ago on December 20, 1937, the ship ran ashore at Princess Royal Island. In December of 1933, the vessel was stranded on the rocks near Anyox. Much farther south on July 23, 1920, she struck a bluff in Seymour Narrows. That time, fortunately, the *Prince George* was able to continue to port under her own steam.

The the most serious accident of her career happened on October 14, 1912. There was fog in Puget Sound. No one saw the halibut boat named the *Lief E*, and the *Prince George* collided with her.

This time, the crew of the *Prince George* was called on to be the rescuers. They lowered lifeboats to the eight fishermen. The fishermen were fine, but little could be done about the 30,000 pounds [13,608 kg] of fish aboard the smaller boat .■

Photo

The Prince George *was involved in five accidents or groundings during its career as a ship for the Canadian National Railway line.*

Winnipeg Under Water

Help Continues to Arrive

MAY 13, 1950

Winnipeg, MB — The water has finally crested. The flooding of the past month has devastated southern Manitoba. An estimated 100,000 people are homeless between Winnipeg and the American border. Recovery will take time, but help continues to arrive from throughout Canada, the USA and Britain.

This week, the Royal Canadian Air Force flew over 1 million pounds [453,600 kg] of supplies to Winnipeg. Supplies also arrived via Trans-Canada Air Lines. The Ontario government has provided flood-fighting equipment and a water chlorinator. The University of Montreal donated typhoid vaccine. Two hundred doctors have provided free services. The Red Cross has sent 80 boats, 50 blankets and 524 hip-waders. The railway has donated the use of box cars, bunk cars and a cook car to nearby Emerson. Even a Lethbridge donut shop sent flood victims 100 dozen donuts.

On April 7, the Premier of Manitoba declared that flooding had reached emergency proportions in the province. However, aid was not forthcoming from the Canadian government. By April 14, the Premier hired the army to come to Manitoba's aid. Initially, 1,000 service men from Princess Patricia Canadian Light Infantry at Calgary and from the 23 Field Squadron at Chilliwack arrived.

Eventually, 1,500 soldiers and volunteers were organized under the direction of Brigadier R.E.A. Morton. The army brought with it 45 bunk beds, 2 water purifiers, 25 life preservers and an amphibious vehicle, nicknamed *The Duck*, to the effort. Although the Legislative Building has been designated Flood Control Headquarters, workers have been everywhere.

The disaster effort has resulted in the largest mass evacuation of people in Canada. Hospital patients and the elderly were sent by train to other areas of rural Manitoba, to Regina and Saskatoon in Saskatchewan and to Fort William and other areas of Ontario. Most evacuees are within a 160-mile radius of Winnipeg, in hospitals and 147 temporary shelters. The flood-control plan went as far as preparing for a mass evacuation of Winnipeg. Now, however, the plan will not be needed.

During the past five

weeks, crews have blasted ice jams, filled sandbags and built dykes, some of which are 30 feet [9 m] high. There has only been one known fatality, a man who was electrocuted by downed power lines. However, issues related to power and clean drinking water remain immediate priorities. Divers from Halifax are tackling many of the underwater problems.

Others have rescued people stranded on the roofs of their homes. RCMP patrol boats, helicopters, rowboats, canoes and *The Duck* have all been used in those rescues.

One of the many individual acts of heroism was performed by a local man. In his boat, he battled currents, gale winds, snapped trees and other debris. As a result, he was able to get 500 loaves of bread to an isolated community.

The flooding began on April 15 after a late spring melt. Last year was wet, and in 1948, there were serious floods in the area. With a high water table, this spring's deluge of wet snow and rain created more serious flood conditions. By May 6, the floodwater had created a lake about 70 miles long, 8 to 12 miles wide and with depths ranging from 6 to 10 feet. Rivers were still on the rise, and only today did the water crest.

Until now, the Canadian government has not had a disaster-relief plan. In fact, the federal government has been the slowest in coming to the aid of Manitoba victims. Despite extensive warnings about the impending disaster, it took no action. One possible preventative measure would have been to open the St. Andrew's locks between Winnipeg and Lake Winnipeg, but Public Works made no such order. Unbelievably, the tax department has not even given Manitobans a break. Despite their homes, belongings and tax receipts being underwater, Manitobans were expected to file returns by April 30 or be penalized.

The Premier plans to negotiate with Ottawa. He wants help with the cost of dykes, fighting the flood and providing immediate relief to victims. ▪

Photo

Winnipeg radio stations have provided around-the-clock flood information. They have broadcast directions to volunteers and conveyed messages to victims' relatives. Since May 7, the Winnipeg Free Press *has been distributed free of charge. It has published maps of affected areas, as well as providing essential flood-relief information.*

Entire Town Comes to Rescue

Shelters over 300 Victims

SEPTEMBER 8, 1952

Juneau, AK — Residents of this northern American port have opened their homes to the crew and 307 stranded passengers of Canada's S.S. *Princess Kathleen*. Captain Hughs has extended thanks to the townspeople, congratulating them for their unbelievably rapid response to the disaster. After the Lena Point shipwreck, buses from Juneau picked up the victims and brought them to town. An hour after they arrived, all had temporary shelter.

On the last cruise of the summer, the Canadian Pacific liner was en route from Skagway to Juneau. It was scheduled to return to port in Victoria. At 2:58 a.m. on September 7, the ship ran aground at Lena Point. Small boats, the US Coast Guard and a steamer named *Alaska* all responded to the S.O.S. wireless radio broadcast.

Although the vessel was not in immediate danger, the captain ordered the lifeboats lowered. About 3-1/2 hours after the accident, all passengers were off the slowly flooding ship. Waiting on shore and enduring windy, cold weather, they huddled in blankets. Some of the ship's crew started fires, heating food and coffee. Others bushwhacked their way to a nearby road.

Once the survivors' exact position was known, buses were sent from Juneau to pick them up. In order to help the stranded passengers, many Alaskans offered to move out of their own homes and stay with friends or relatives.

Currently, the CPR is making plans to have passengers returned to Canada. The company has arranged for a DC-3 airplane and two float planes to assist those needing immediate transportation to Vancouver. The *Princess Elizabeth* has set sail to pick up other stranded passengers.

Although there was no loss of life when the *Princess Kathleen* grounded, the ship was seriously damaged. The captain and some crew members remained onboard trying to save the twenty-seven-year-old vessel. With the rising tide, pumping operations became futile. By 11:30 last night, those still aboard were forced to abandon ship.

At about 2:00 this morning, the stern became submerged in 130 feet [40 m] of water. The bow is under about 60 feet [18 m] of water. Given the problems of salvage in these northern waters, the $5 million ship may never be raised.

This is the second accident in little more than a

year for the *Princess Kathleen*. On August 3, 1951, the *Kathleen* and the *Prince Rupert* were sailing in fog and rain. On both ships, the captains had retired for the night and second officers were in charge. Both had radar, a reasonably new and still somewhat unfamiliar tool in ship navigation. At about 4:30 a.m., visual contact was lost. After the collision, both ships were badly damaged. However, they made it back to port under their own steam. ■

Photo

Here, the Princess Kathleen *is in drydock for $250,000 worth of repairs. They were required after last year's collision with the CNR ship, the* Prince Rupert.

Mountain Claims Lives of Three Women

Guide Also Perishes in Tragic Fall

JULY 31, 1954

Banff, AB — A party of eight Mexican climbers, seven women and a male guide, came to climb in the Rockies. Tragically, only four women will return safely to their homeland. Yesterday, a fall took the lives of the guide and three women. Late last night, three others were rescued from Mount Victoria by Canadian climbers. The fourth had spent the day at the Abbot Pass Climbers' Hut.

On the morning of July 30 at the Chateau Lake Louise, mountain guide Walter Feuz spoke to the climbers. On June 29, the Mexican group camped near the teahouse on the Plain of Six Glaciers. Their guide, Eduardo Sanvicente, was a very experienced climber in other areas of the world but not in the Rockies. On the day of the tragedy, all intended to go as far as Abbot Pass Hut. Then, six of the women and their guide planned to set out for the south peak of Mt. Victoria, towering behind Lake Louise. They planned to use ropes, with three women on one and the guide and three others on the other. By noon, they hoped to reach the peak.

That afternoon, Walter Feuz looked though his binoculars to check their progress. Only three women were still visible on the mountain. They were not moving and were in a dangerous position on the snow face. Assuming that the others had fallen, he raised an alarm.

His brother, Ernest, also a renowned mountain guide, led the rescue party. Since pack horses were not immediately available, Walter boated his brother, Charles Rowland and three other climbers to the far end of Lake Victoria. From there they continued on foot. Walter was to bring the horses as soon as possible so that once the women were rescued, they could be brought back to the chalet.

While climbing, the rescue party came upon the bodies of four fallen climbers. All were beyond help, so the rescuers continued toward the survivors. At 6:40 p.m., they reached the Abbot Pass Hut. There, one Mexican woman had remained to cook supper for the others. Because she spoke only Spanish, no one could communicate the tragic circumstances of her

friends.

From there, Feuz and Rowland continued up the mountain while the others prepared a fire, blankets, tea and soup for the survivors. In 35 minutes, the two climbers reached a ridge above of the stranded women. The men tied two 150-foot [46 m] ropes together. Using his axe for a belay, Feuz was to be the anchor and pull the survivors to the ridge. Rowland dropped down to them. They were too cold and terrified to move. Finally, one by one, he managed to tie the rope around them and Feuz pulled them to safety.

Once the men had the survivors off the snow face, it was another two hours before the party reached the climbers' cabin. Although warmed with blankets, hot drinks and food, the women were not given time to think about their tragedy. Wanting to take them past the bodies of their friends while it was still dark, rescuers roped themselves and the women on three-man ropes and continued down the mountain.

The group left the cabin shortly before 11:00 p.m. With only two flashlights between them, they walked for two hours on the dark trail. At 1:00 a.m., they arrived at the teahouse. There, in the small staff cabin nearby, the women and two of the rescuers had shelter for the rest of the night.

The women's tragic story was simple. They had reached the peak and began descending. A woman on the guide's rope slipped. Unable to break the fall, he and three of the women fell 2,000 feet [610 m] to their deaths. After the fall, the three survivors started to descend. Terror had paralysed them and they crouched on the snow until they were rescued. ■

Bottom Photo

The Mexican climbers traveled up the eastern snow face rather than the commonly used ridge route.

Top Photo

Six women, ranging in age from 21 to 35, were involved in the tragedy on Mount Victoria. This photo was taken at the top of the south peak before they began the fateful descent. All but one of the women had considerable experience climbing in Mexico. Their trip had been organized by the Alpine Club of Mexico.

Mountain Rescue Critical

Special People Needed

SEPTEMBER 5, 1955

Calgary, AB — The Mountain Search and Rescue Team for the Canadian Parks has experienced one of its most challenging years. Members of the team are well trained and extremely skilled in climbing, mountaineering and first aid, but they have had to call on special internal resources and reserves this summer.

Most recently, they had the difficult job of recovering the body of Frank Koch. On September 3, 1955, the Vancouver climber fell while on Mount Blane, a mountain about 50 miles southwest of Calgary in the Kananaskis area. Climbers consider the 9,600-foot [2,926 m] peak to be treacherous. They found Koch, 23, at the 8,000-foot [2,438 m] level. The search-and-rescue team then faced the steep descent. Although the face was perpendicular, they were successful.

Koch was the tenth climber to die in the Rockies during the season. On June 13, 1955, seven teenage boys from Philadelphia died while climbing Mount Temple. Four survived the avalanche that swept down on the group. Their adult leaders were not with them at the time. Walter Perren and Chief Warden Bert Pittaway were involved in the search and rescue.■

Photo

Mountain Climbing 100, Mt. Blain Sept 3, 1955. Here, the mountain rescue team, led by Assistant Chief Warden Walter Perren of the Parks Mountaineering Service, brings Koch's body from Mount Blane.

Man Survives Six Days

Owes Life to Friend and Boy Scout Lesson

SEPTEMBER 5, 1955

Vancouver, BC — Alick Patterson has finally been rescued from Mount Seymour. He survived six days and nights without food.

On Saturday, October 20th, the twenty-four-year-old was found on the 4,000-metre mountain which overlooks Vancouver's harbour. He was flown off the mountain in a Royal Canadian Air Force helicopter. Although hospitalized, he is in good spirits and doctors expect a full recovery. One companion, twenty-three-year-old Robert Duncan, found his way to safety. Another, Gordon McFarlane, lost his life.

When the three Scottish immigrants were reported missing, a search-and-rescue mission was initiated. Along with local climbers, RCAF helicopters participated in the search. However, they had no success until Duncan found his way off the mountain. When he described the general location of Patterson, RCMP and 30 B.C. mountaineers, mostly from the Alpine Club, combed the western slope.

For the young men, the ordeal began with what they assumed would be a short walk down from Mount Seymour Ski Lodge. They had no food or gear. Fog settled around them and the group became lost. They found a cave in which they kept dry and reasonably warm the first night. The next day, while trying to find their way down, they followed a stream. They came across an overhang which again protected them from the rain. Duncan and Patterson remained there until morning.

McFarlane, who was twenty-seven, was impatient and decided to continue without them. The next day, Patterson was still exhausted. Although he is a towering young man, he weighs only 120 pounds [54 kg]. Duncan is the huskier of the two, and Patterson convinced him to set out to try to save them both.

Duncan was on the fog-shrouded mountain for three days before he found his way off the mountain. In the meantime, he had discovered McFarlane's body. The twenty-seven-year-old had fallen and drowned. Fortunately, Duncan was able to give rescuers enough information to find his friend Alick Patterson. ■

62 Perish in Plane Crash

Bodies Not Recovered Until Spring

MAY 14, 1957

Chilliwack, BC — Searchers and crash-site investigators are in agreement that no more can be done until after the spring melt. The location of the 62 victims in the crash of a Trans Canada Airline [TCA] plane has been found. Until recovery work proceeds in July, approaches to the area have been sealed by RCMP.

The passenger plane went missing in foul weather on December 9, 1956. The plane, presumed down in the Skagit Valley, has been found on Mount Slesse, less than 100 miles from Vancouver. To date, the crash is Canada's worst air disaster.

When rescue parties could find no evidence of the crash, the search was suspended. The Royal Canadian Air Force and TCA planned to resume aerial searches later this year. Two days ago, climbers found the plane. It was near the 7,500-foot [2,286 m] level of Mount Slesse. B.C.'s renowned woman climber, Elfrida Pigou, brought the first piece of wreckage down from the mountain.

Once the metal was positively identified, a helicopter air search located the wreckage. Immediately, the RCAF Rescue Co-ordination Centre and RCMP became involved in the search. No survivors could be expected, but members of the Mountain Rescue Group and other climbers, including Pigou, went by helicopter to the mountain.

Although they approached the site of the crash, to go closer was to risk their lives. With loose rock and a vertical drop of 2,500 feet [762 m] to the snow, they determined that the recovery of the bodies should wait until the snow melts. The plane was so fragmented that debris was found in crevasses lower on the mountain. Even in the summer, work will be difficult. ■

Photo

Flying in the mountains in winter is a challenge. This small plane crashed near the Sunshine Ski Chalet. Former RCAF pilot Al Gaetz and Richard Pike were flying supplies in to skiers at Mount Assiniboine. It was cold, but visibility was good. On the way, Pike wanted to get an aerial photo of Sunshine Lodge. Probably overloaded, the plane lost power, headed straight for the rocks of Strawberry Hill and crashed. The gas tank ruptured and Gaetz was covered in gas.
The two escaped through the window and floundered in the snow to get as far as possible from the aircraft. Fortunately, there was no fire. People from the lodge quickly skied to their rescue. The young men had only minor injuries, and they were able to ski away from the wreck and to the lodge on their own.

Planes and Whirlybirds at Work

New Era for Police, Fire and Rescue

MAY 26, 1957

Dawson City, YK — A helicopter has gone to the rescue of two men stranded without food as a result of flooding in the Yukon. The men were reported to be 90 miles [145 km] south of Dawson City. In many isolated areas affected by floodwaters, the Yukon and Klondike Rivers are 43 feet [13 m] above normal.

The military has flown sandbags and water pumps to Dawson from its base in Edmonton. To protect the northern city from flooding, residents and military personnel have filled 14,000 sandbags in 36 hours. Additional pumps and sandbags will be flown to Whitehorse if they are needed there.

In some of the flooded areas, helicopters offer the only access and means of help or rescue. Although they are a new addition to disaster-relief programs, helicopters have the advantage of being able to land in conditions and on terrain where runway planes and float planes can't land. They are speedy and flexible, but have their limitations. They can't compete with air-transport planes when it comes to carrying heavy loads of food, clothing and equipment to disaster areas.

The helicopter assistance to Dawson during the flood is courtesy of the Royal Canadian Air Force. The RCMP is also considering the usefulness of the whirlybirds in fighting crime. Since 1919, air service has been considered for Mounties' use.

Beating them to the draw that year, Detective James Campbell of Edmonton's city police department illustrated the usefulness of planes in finding criminals. He was the first Canadian policeman to board an aircraft as part of a manhunt. The plane, piloted by Captain Wilfrid [Wop] May, was used to search the wilderness for the murderer of a city policeman. The suspect was an ex-convict. When both Campbell and former Royal Air Force officer May stepped into the plane, they were armed. Good at their jobs, they found and apprehended their suspect.

In the years that followed, the Mounties also used air transportation

in their jobs and began using planes for various duties in 1921. The RCMP depended on military and commercial or privately owned air transportation. Since 1932, RCMP have conducted aerial patrols along the coasts. Throughout their investigations, they were provided with air force planes. Finally, in 1937, the RCMP bought four planes to patrol Canada's coasts and prevent smuggling.

Now, the helicopter is demonstrating its potential usefulness. With its proven success in manoeuvring in areas inaccessible to other aircraft, there may be more than disaster relief in store for the whirlybirds. ▪

Photo

This Bell 47 helicopter is being used by the Coast Guard for search, rescue and patrol duties near Victoria.

Bridge Collapse Claims Eight

Heroic Efforts Save Many More

JUNE 18, 1958

Vancouver, BC — The collapse of the Second Narrows Bridge at 3:40 p.m. yesterday is presumed to have taken 18 lives. About twenty people are in hospital. Others were rescued or managed to scramble to safety without serious injury.

In the terrible tragedy of Black Tuesday, many also became heros. Men screamed for help and others risked their lives for the victims.

Art Pirlon and Loyd McAtee rescued seven men who were injured and in the water. Both were working at the north end of the bridge on a survey crew. When the bridge collapsed, they jumped into a rowboat. As the tide was coming, Pirlon grabbed everyone he could see. Said Pirlon, "I ain't no hero. Everyone was pulling guys out. I don't want to talk. My own brother…my own brother…is down there somewhere."

Six welders were working a couple of blocks away. They threw equipment in a truck, careened to the scene and jumped in a boat. Immediately, they began the grisly work of cutting trapped men from the tangle of steel beams. One man had crawled as far as possible up a beam, but he was still chest-deep in water and the tide was coming in. The welders fired their torches and piece by piece, they removed the steel cage that trapped him. They released another seven men from their prisons. Of those men, six died.

Another hero was John Wolf, aged 28, who was directing painters on the bridge. Seeing men in trouble, he made his way to a rope dangling from a beam. He slid down it, and when his hands were raw from rope burn, he dropped to rescue a victim.

Another worker grabbed two men in the water. In the end, he could only hold on to one. For half an hour, until a boat came to help, the rescuer held on to the semi-conscious man who could never have saved himself.

One rescuer was working on a speedboat that belonged to a businessman. When the workman's wife told him of the calamity, he "borrowed" the boat and saved six lives. He pulled injured people from the water while those on shore directed him to others. In the process the boat suffered, but men lived.

The Workmen's Compensation Board had ruled that everyone assigned to a job over the water must wear a life jacket. One survivor was

attached to a beam by his safety harness. When the span fell, he dropped 30 feet. The sinking beam dragged him into the water. Instead of succumbing to panic, he was able to unhook his harness. His bright yellow life jacket brought him to the surface where, like others, he was rescued.

Men on 60 small boats rushed to help. Some had grisly work. The captain and crew of one boat discovered debris and bodies floating toward them. The dead victims were very badly cut, but the would-be rescuers pulled in men who were past help, thinking of the families who would never know comfort if their loved ones were lost at sea. The job was more distressing than anything they had ever faced before.

Diver Don Sorte heard of the disaster and grabbed his diving equipment. He joined other volunteers to search the depths. Given the highest tide in June, the treacherous current under the bridge and the wash of boats, the work was extremely dangerous. As a result, many of the frogmen began their work only today. In the murky water, visibility is limited. Still, they search.

The city's ambulance workers have rushed between the scene and the emergency wards of the Vancouver and North Vancouver general hospitals. There, staff continues to care for the injured. Police are assigned to guard the bridge and prevent distraught families and curious onlookers from further disaster. ▪

Photo

The Second Narrows Bridge was under construction when two spans crumpled into about 200 feet [61 m] of water. On those spans were 60 iron workers and 24 painters. As yet, the reason for the collapse is unknown. The disaster is Vancouver's worst since the fire of 1886.

Smoke Jumpers, Smoke Eaters and Other Heros

Alberta Government Moves to the Modern Age of Forest Fire Fighting

JANUARY 1, 1958

Edmonton, AB — The Alberta government is entering the modern age of forest fire fighting. It plans to purchase its first airplane and helicopter intended solely for forest fire prevention and suppression.

Aerial fire fighting techniques have been implemented for decades. On September 9, 1950, Ontario became the first province in Canada to waterbomb a forest fire from the air. Dropping special paper bags full of water from a DeHavilland Beaver, the province's forestry service discovered that their method achieved results and had potential. By 1957, Ontario was already experimenting by dropping special chemical fire retardants from air tankers.

Americans in Washington State have experimented with that possibility since 1930. However, 90 per cent of Canadian forests are within five miles of a lake. These lakes provide water for suppression and for the landing of ground-suppression teams. As a result, Canadian provinces will continue to take advantage of that inexpensive fire-fighting resource.

Modern methods of fighting forest fires require complex planning. A well-informed, well-trained and disciplined team is also essential. If the fire is a large one, the order of attack becomes very important to the protection and safety of the men going into a fire site.

First, an air-attack boss is flown to the site. From a position in the air or on the ground, he plans the operation. Then an air-tanker boss is sent to the scene for planning purposes. Later, the first-attack pilot takes smoke jumpers or fire-fighting cargo. Smoke jumpers, cargo drops and helicopter operations do not happen simultaneously because of the danger to men and aircraft.

Smoke jumpers are well-trained men who are prepared to work alone in

remote areas. They take initial action to suppress large fires and extinguish small ones. Smoke chasers do similar work. Some travel in vehicles and others on aerial fire-detection patrols. However, those on aerial patrol expect to be landed near the site of the fire.

In some areas of the country, helijumpers are being added to fire fighting teams. These jumps are dangerous for both jumper and helicopter pilot. However, such daredevils perform a valuable role fighting fires. They enter terrain where parachuting from high-flying planes is too dangerous and aircraft landing is not possible.

The helicopter pilot and jumper search for a location that is as flat and open as possible. They make at least four passes. The high reconnaissance pass is to select a general site. A low pass determines the specific site. Another low one is made to drop equipment, and the lowest one is for the jump.

In this dangerous work, the helicopter pilot must hover 10 feet above the jump zone at a reduced speed of 10 miles [16 km] per hour or less. In such close proximity to a wildfire, the risk is high. To these men, however, the forests are worth the risk. ▪

◦ *Photo*

Saskatchewan has a long history of using airplanes to protect and save its forests. Here, a Saskatchewan smoke jumper parachutes into the dangerous world of the forest fire. For hundreds of years, other firemen have proudly claimed the nickname "smoke eaters."

Bibliography

Anderson, Frank. *Regina's Terrible Tornado.* Aldergrove, BC: Frontier Publishing, Ltd., 1968.

Anderson, Frank. *Canada's Worst Mine Disaster.* Aldergrove, BC: Frontier Publishing, Ltd., 1969.

Anderson, Frank. *The Rum Runners.* Edmonton, AB: Lone Pine Publishing, 1991.

Appleton, Thomas. *Usque ad Mare.* Ottawa, ON: Department of Transport, 1968.

Baird, Donal. *The Story of Firefighting in Canada.* Erin, ON: The Boston Mills Press, 1986.

Basque, Garnet, (editor). *Frontier Days in Alberta.* Langley, BC: Sunfire Publications, Ltd., 1992.

Bell, Ken and C.P. Stacey. *100 Years: The Royal Canadian Regiment, 1883-1983.* Don Mills, ON. Collier Macmillan Canada, Inc., 1983.

Bondar, Barry. *Vancouver: The Story and the Sights.* North Vancouver, BC: Whitecap Books, nd.

Bumsted, J.M. *The Manitoba Flood of 1950.* Winnipeg: Watson and Dwyer Publishers, 1993.

Canadian Alpine Journal 22 (1933); 29 (May, 1945); 38 (May, 1955); 40 (May, 1957).

Canadian Encyclopedia. James H. Marsh (editor-in-chief). Edmonton: Hurtig Publishers, 1985.

Chronicle of Canada. Elizabeth Abbot (editor-in-chief). Montreal: Chronicle Publications, 1990.

Cohen, Stan. *The Streets Were Paved With Gold.* Revised Edition. Missoula, MN: Pictorial Histories Publishing Co., 1988.

Delaney, William et al. *Saskatoon: A Century in Pictures.* Saskatoon, SK: Western Producer Prairie Books, 1982.

Denny, Cecil. *March of the Mounties.* Surrey, BC: Heritage House Publications, 1994.

Dowling, Phil. *The Mountaineers.* Canmore, AB: Coyote Books, 1995.

Dobrowolsky, Helene. *Law of the Yukon.* Whitehorse, YK: Loose Moose, the Yukon Publishers, 1995.

Fairley, Bruce. *The Canadian Mountaineering Anthology.* Vancouver, BC: Lone Pine Publishing, 1994.

Gilkes, Margaret and Marilyn Symons. *Calgary's Finest.* Calgary, AB: Century Calgary Publications, 1975.

Gould, Jan. *Women of British Columbia.* Saanichton, BC: Hancock House Publishers, Ltd., 1975.

Gourley, Catherine. *Island in the Creek.* Madeira Park, BC: Harbour Publishing, 1988.

Hacking, Norman. *Prince Ships of Northern B.C.* Surrey, BC: Heritage House Publishing Co., 1995.

History of the Canadian West. Vol 2 (1982), Vol 3 (1984). T.W. Paterson (editor). Langley, BC: Sunfire Publications.

Horrall, S.W. *The Pictorial History of the Royal Canadian Mounted Police.* Toronto/Montreal: McGraw-Hill Ryerson Ltd., 1973.

Kauffman, Andrew and William Putnam. *The Guiding Spirit.* Revelstoke, BC: Footprint Publishing, 1986.

Kelly, William and Nora. *The Horses of the Royal Canadian Mounted Police.* Toronto: Doubleday Canada, Ltd., 1984.

Kerr, Don and Stan Hanson. *Saskatoon, SK: The First Half-Century.* Edmonton: NeWest Publishers, 1982.

Bibliography

Linkewich, Alexander. *Air Attack on Forest Fires.* Altona, MB: Printed by D. W. Friesen & Sons, 1972.

Klondike Letters: The Correspondence of a Gold Seeker in 1898. Juliette Reinicker (editor). Anchorage, Alaska: Alaska Northwest Publishing Co., 1984.

Klucker, Michael. *Vancouver: The Way It Was.* Vancouver, BC: White Cap Books, 1986.

Knuckle, Robert. *In the Line of Duty.* Burnstown, ON: General Store Publishing House, 1994.

Leavitt, Clyde. *Forest Fire Prevention in Canada.* Toronto, ON: Bryant Press, 1912, 1913.

Lothian, W.F. *A Brief History of Canada's National Parks.* Ottawa: Environment Canada, Parks Canada, 1987.

Mair, A.J. *E.P.S.; The First 100 Years.* Edmonton, AB: Edmonton Police Service, 1992.

Mactavish, J.S. and M.R. Lockman. *Forest Fire Losses in Canada.* Ottawa, ON: Department of Forestry, 1961.

MacDonald, Robert. *Making Vancouver, 1863-1913.* Vancouver, BC: UBC Press, 1996.

McClement, Fred. *The Flaming Forests.* Toronto/Montreal: McClelland and Stewart, Ltd., 1969.

Nicol, Eric. *Vancouver.* Toronto: Doubleday Canada, Ltd., 1978.

Outlaws and Lawmen of Western Canada. Vol 1, Vol 2, Vol 3. Surrey, BC: Heritage House Publishing Co., 1992.

Parker, Davis. *First Water, Tigers!* Victoria, BC: Sono Nis Press, 1987.

Partners. Calgary, AB: Calgary Police Service, 1987.

Pethick, Derek. *British Columbia Disasters.* Langley, BC: Stagecoach Publishing Co., 1978.

Pethick, Derek. *Victoria: The Fort.* Vancouver, BC: Mitchell Press, 1968.

Pioneer Policing in Southern Alberta. William Baker (editor). Calgary, AB: Historical Society of Alberta, 1993.

Richardson, A.H. *Forestry Chronicle.* Vol IX (June, 1933); Vol XI, (December, 1935).

Sandford, R.W. *The Canadian Alps.* Banff, AB: Altitude Publishing, 1990.

Tragedies of the Crowsnest Pass. Surrey, BC: Heritage House Publishing Co., Ltd., 1983.

Turner, Robert. *The Pacific Princesses.* Victoria, BC: Sono Nis Press, 1977.

Original Sources

The Calgary Herald. (Calgary, AB) 2 January, 1924, 6 January, 1913; 4 July, 1924; 25 April, 1929; 3 September, 1955; 20 October, 1956; 22 October, 1956; 27 October, 1957.

The Edmonton Journal. (Edmonton, AB) 28 June, 1915; 21 September, 1940.

The Lake Windermere Echo. Supplement, 1985.

The Manitoba Daily Press. (Winnipeg, MB) 12 April, 1893; 14 April, 1893; 21 April, 1893; 28 April, 1893; 3 May, 1893; 5 May, 1893.

The Morning Leader. (Regina, SK) 8 June, 1908; 10 July, 1908; 5 March, 1912. *The Leader Post* (Regina, SK), 1 July, 1929; 1 July, 1935; 2 July, 1935.

The Vancouver Sun. (Vancouver, BC) 18 June, 1958.

Photo Credits

Western Canada Pictorial Index: 95

BC Archives & Records Service: 7, 9, 29, 39, 45, 75, 86, 93, 97

Vancouver Public Library: 15, 107

City of Victoria Archives: 21A, 21B

Revelstoke Museum & Archives: 47, 73

City of Vancouver Archives: 61, 91

Provincial Archives of Alberta: 11, 51, 57, 65

Glenbow Archives: Front cover, 17, 25A, 25B, 31, 35, 37, 43, 55, 69A, 69B, 71, 77, 99B, 100

Glenbow Museum: 85B

Whyte Museum of the Canadian Rockies: 23, 67, 99A, 103

City of Edmonton Archives: 59, 89

Edmonton Police Museum Archives: 81

Calgary Police Service Archives: 85A

City of Lethbridge Archives & Records Management: 33

Saskatchewan Archives Board: Back cover, 41A, 41B, 49, 53, 79, 83, 109

RCMP Archives, Regina: 63

Provincial Archives of Manitoba: 19, 65

Yukon Archives/University of Washington Collection: 27

Courtesy of Faye Holt: 112

About the Author

Faye Reineberg Holt is a Calgary researcher, writer and freelance editor. She has had extensive experience teaching high school and has offered writing-related workshops to people ranging in age from elementary school to senior citizens.

Her first full-length book, *Alberta: A History in Photographs,* was published by Altitude in 1996. She has also published a poetry chapbook called *Ice Fog,* as well as many poems, short stories and articles. Faye is an avid hiker and gardener. She also enjoys cooking, skiing and traveling.

This book is dedicated to her husband, Walt, and her three sons, Tobin, Mark and Graham, from whom she has learned something about perseverance, insight and courage—the qualities found in heros.